IET

The Institution of Engineering and Technology

Guide to

Implementing Electrified Heat in Domestic Properties

Publication information

Published by The Institution of Engineering and Technology, London, United Kingdom

The Institution of Engineering and Technology is registered as a Charity in England & Wales (no 211014) and Scotland (no SC038698).

IET

© 2022 The Institution of Engineering and Technology

First published 2022 (ISBN 978-1-83953-392-1)

Copies of this publication may be obtained from:
PO Box 96
Stevenage
SG1 2SD, UK
Tel: +44 (0)1438 767328
Email: sales@theiet.org
theiet.org/electrical

ISBN 978-1-83953-392-1 (paperback)
ISBN 978-1-83953-393-8 (electronic)

Typeset in India by MPS Limited
Printed in the UK by Hobbs the Printers Ltd, Brunel road, TOTTON, Hampshire, SO40 3WX

Contents

Contents

Contents

▆ Acknowledgements

The Institution of Engineering and Technology (IET) acknowledges the contribution made by the following individuals and organizations in the development of this document:

Technical Author
Paul Bennett

Technical Committee
Max Halliwell
Edmund Hunt
Sam Hunt
Shaun Hurworth
Craig Pilkington
Robert Sansom
Cameron Steel
Geoff Smyth
Oliver Sutton

Organizations
BEIS, BSSEC, Carbon Trust, Energy Systems Catapult, Glen Dimplex Heating & Ventilation, Mitsubishi Electric, Silver EMS, SMS PLC, WSP

Special thanks to
Dr Richard Jack (Build Test Solutions Ltd) for his assistance in the approach to Section 2.7 and Figure 2.8 and Gabor Szabo for his help with the schematics included in Appendix D.

Scope and purpose

This Guide has been prepared for all those involved in the design, selection, installation and operation of direct electric heating and heat pumps in new and existing domestic properties in the UK. It also provides an insight into market models of how heat may be provided in the future.

▀▄ Foreword

Climate change is perhaps the greatest challenge of our time. Not just for the public, but for policy makers, manufacturers, installers and all other stakeholders in the energy system.

The UK has committed to mitigate climate change by reducing its emissions to net zero by 2050. To achieve this, almost every aspect of the energy system will need to change and emissions from buildings will likely need to be almost eliminated.

The challenge for buildings is largely associated with how we heat the space and water for our everyday use. Space and water heating in buildings accounts for approximately 23 % of all UK greenhouse gas emissions[1]. Of these emissions, the largest contributors are from fossil gas and oil boilers, which have been commonplace in UK buildings for decades. If we are going to meet our targets, these heating systems will need to change, and the UK is committed to introducing ambitious long-term policies to enable the market to respond to this opportunity.

One of the key options for this change is to switch to electrified heating. Electric heating technologies have been around for over 100 years and supply us with most of our heating needs. They can vary from simple electric panel radiators to more complex and highly efficient heat pumps. Crucially, because they are powered by electricity, they all hold the potential to be decarbonized, alongside the decarbonization of the power sector.

This is an immense challenge, and it is vital that knowledge and skills are disseminated as widely and as quickly as possible if we are to have the workforce we require to make this transition.

This Guide provides an excellent introduction to the challenges and solutions available to experts wanting to understand a transition to electrified heating.

Oliver Sutton
Department for Business, Energy and Industrial Strategy (BEIS)

---

[1]Department for Business, Energy and Industrial Strategy *Final UK greenhouse gas emissions national statistics: 1990 to 2018* BEIS, 2020. Available at: https://www.gov.uk/government/statistics/final-uk-greenhouse-gas-emissions-national-statistics-1990-to-2018

≡ Section 1

Introduction

This Section discusses the history of household heating in the UK and the widescale adoption of natural gas. It explores the impact of gas consumption on greenhouse gas emissions and the need to transition to a low-carbon form of heating to avoid the catastrophic effects of climate change. The demand for heating is explored and the main factors that affect future demand, such as housing insulation but also milder winters from climate change itself. Options for low carbon heating are then discussed with the focus on electric heating solutions. Whilst this Guide is directed towards the electrification of heat, many of the principles discussed apply equally to other future heat options.

1.1 How do we heat homes in the UK?

For most of the first half of the twentieth century, coal dominated UK primary energy consumption[2]. It was used to produce electricity, transport (railways, shipping) and town gas and was also used to heat buildings, such as the houses shown in Figure 1.1.

Figure 1.1 Twentieth century housing in Aberdeen

After the Second World War, consumption of coal declined because of the growth in other energy sources such as oil and nuclear power. With the advent of North Sea natural gas in the 1960s, coal for domestic heating consumption fell rapidly and this decrease accelerated with the growth in natural gas for household heating (see Figure 1.2).

Natural gas transformed space and water heating in domestic households with the widescale deployment of central heating. This facilitated whole-house heating, with locally controlled radiators with thermostatic radiator valves (TRV) as shown in Figure 1.3, which improved comfort levels and reduced the health hazards associated with cold and damp buildings. By 2015, natural gas for domestic heating was being used in over 85 % of UK households (predominantly those with central heating systems), with around 6 % of households being heated by electricity and the remainder by oil and bioenergy (see Figure 1.4).

---

[2]Department for Energy and Climate Change (DECC) *Digest of United Kingdom energy statistics 60th anniversary* DECC, London, 2009. Available at: https://www.gov.uk/government/statistics/digest-of-uk-energy-statistics-dukes-60th-anniversary

Section 1 – Introduction

Figure 1.2 Domestic consumption of fuel in the UK (1970–2020) showing the transition from heavy coal usage to an increase in the national uptake in natural gas

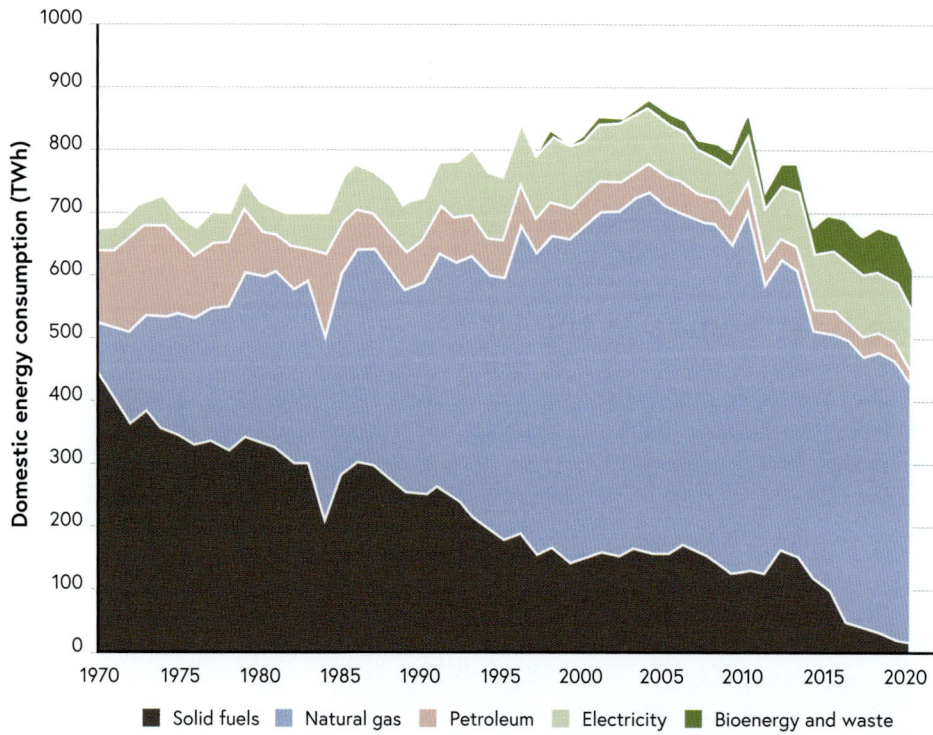

Figure 1.3 Typical domestic gas central heating radiator

Section 1 – Introduction

Figure 1.4 UK-installed central heating systems by type showing the considerable uptake in heating systems powered by natural gas

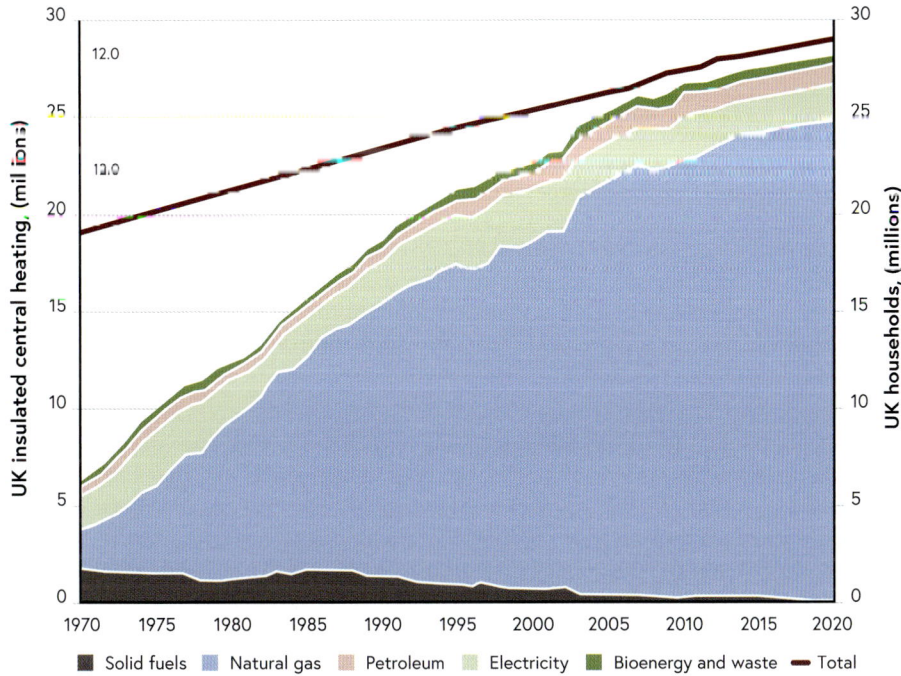

1.2 Why the UK wants to stop using natural gas for heating

Natural gas collected from offshore drilling platforms, particularly in the North Sea (see Figure 1.5), offers many advantages to households. It is clean, easy to use and compared with other forms of heating, economic. Modern condensing gas boilers can operate at 90 % efficiency although their actual efficiency depends on the heat load, and can also provide hot water on demand, thereby avoiding the need for hot water storage. Typically, gas boilers have a heat output significantly more than the normal building requirements, which means

Figure 1.5 Offshore North Sea drilling platforms

they can be very responsive to heating demands. However, the drawback is that when natural gas is combusted it produces CO_2, which is a greenhouse gas (GHG) and the main anthropogenic contributor to climate change. To avoid the associated catastrophic impacts, the UK government has committed to reducing its GHG emissions to net zero by 2050 and has stated in its energy white paper published in December 2020[3] that the UK will have to transition away entirely from traditional natural gas appliances for heating.

1.3   How climate change is affecting the UK

In 2018 the UK's Met Office published its second report (UKCP18) that provides the most up-to-date assessment of how the UK climate may change in the future[4]. The assessment explores the impact of global GHG emissions and their impact on the UK. It concludes that there will be a greater chance of warmer, wetter winters and hotter, drier summers with more frequent and intense extreme weather events, such as heatwaves and heavy rain.

Two climate change outcomes are presented based on an increase in global mean temperatures of +2 °C and +4 °C above pre-industrial levels. These show that the UK could experience average temperature increases of up to +6 °C in the summer and +4 °C in the winter. However, if the global mean can be kept to +2 °C above pre-industrial levels then summer and winter temperature increases are expected to be less, that is, +4 °C and +3 °C, respectively. Even so, the impact on the UK's weather is likely to be profound. The warmer winters will mean a reduction in the heat required for household space heating by an estimated 20 %. This may have implications for the sizing of heat systems. Additionally, the hotter summers might result in the need to incorporate cooling systems into buildings with passive measures not involving mechanical or electrical systems, such as building adaption and design as well as ventilation with and without cooling.

1.4   Climate Change Act 2008 (and 2019 amendment)

In 2008 the UK government passed the Climate Change Act 2008. This legislation required an 80 % reduction in UK GHG emissions by 2050 compared with emission levels in 1990. It also established the Committee on Climate Change (CCC) to advise the government and report on progress. In 2019 the Act was amended – The Climate Change Act 2008 (2050 Target Amendment) Order 2019 – and the 2050 target changed from an 80 % reduction in emissions to a 100 % reduction. This is frequently referred to as a 'net zero' target as it is recognized that there are GHG emitting sectors that are at present hard to decarbonize, for example, aviation and shipping. Therefore, to achieve zero GHG emissions these sectors will need to be 'offset' by negative emissions from elsewhere, such as technologies that capture CO_2 from biomass energy production and planting trees.

1.5   What are the UK's CO_2 emissions?

The decline in the UK's CO_2 emissions since 2000 has been substantial. This is mostly due to the reduction in coal consumption, but natural gas remains a dominant source of the UK's CO_2 emissions. As over 85 % of UK households use natural gas for heating, with an average CO_2 emission of 3 tonnes per year, the contribution to the UK's CO_2 emissions is substantial. Natural gas is also used for power generation for industry and for commercial buildings, but the domestic sector contributes the most to CO_2 emissions. Figure 1.6 presents the UK's CO_2 emissions since 2000.

---

[3]Department for Business, Energy and Industrial Strategy (BEIS) *Energy white paper: Powering our net zero future* BEIS, London, 2020. Available at: www.gov.uk/government/publications/energy-white-paper-powering-our-net-zero-future
[4]Met Office *UK Climate Projections* Available at: www.metoffice.gov.uk/research/approach/collaboration/ukcp/index

Figure 1.6   UK CO_2 emissions

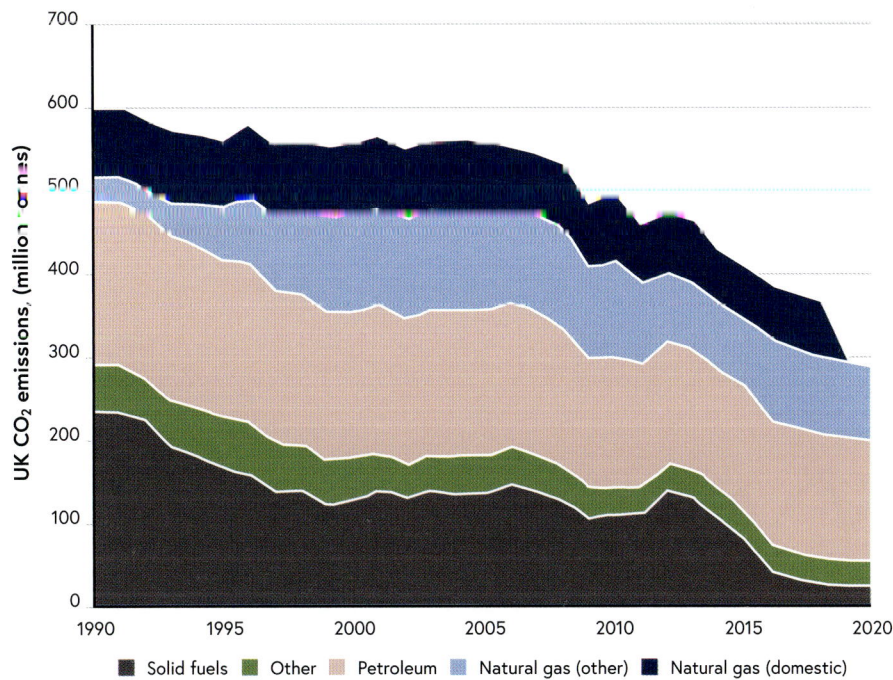

1.6   The factors determining the demand for heat

Water heat has slowly declined in consumption from around 5 MWh per year to just over 2 MWh per year. This reduction has probably been driven by more efficient hot water storage systems and washing appliances, greater use of electric showers and the use of lower temperatures for washing clothes but, most importantly, the installation of combination gas boilers that heat water on demand.

Since 2000, combination (or combi) gas boilers, such as the example shown in Figure 1.7, have increased from around four million to just under 15 million in 2017. At the same time the number of hot water

Figure 1.7   Domestic combi gas boiler

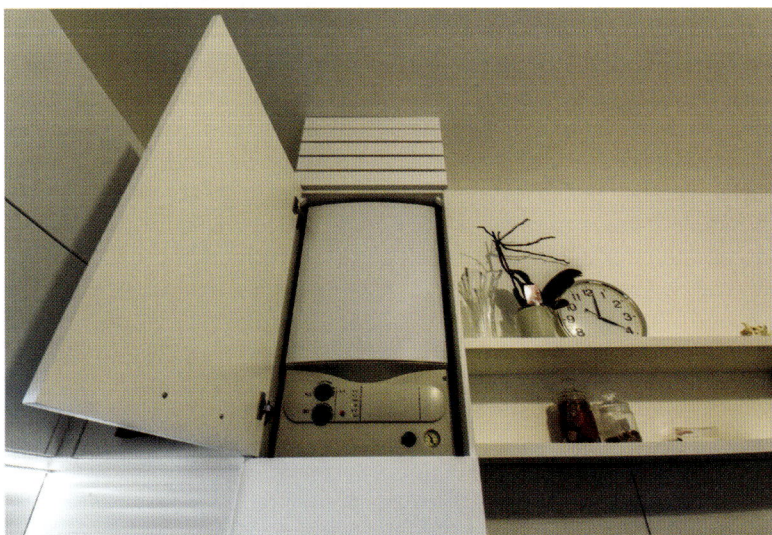

Section 1 – Introduction

storage cylinders have reduced from 20 million in 2000 to 11 million in 2017 (based on the assumption that hot water cylinders are removed when combi gas boilers are installed thereby releasing valuable household space). Combi gas boilers have sufficient power capacity to supply hot water on demand for personal washing purposes, around 20–30 kW_{th}. Electric water heating on demand would not be possible for a typical domestic household without an upgrade of the network connection capacity and therefore it is likely to involve retrofitting some form of hot water storage system within the household.

Figure 1.8 shows the UK average household demand for heating and hot water since 1970.

Figure 1.8   UK domestic demand for heating and hot water

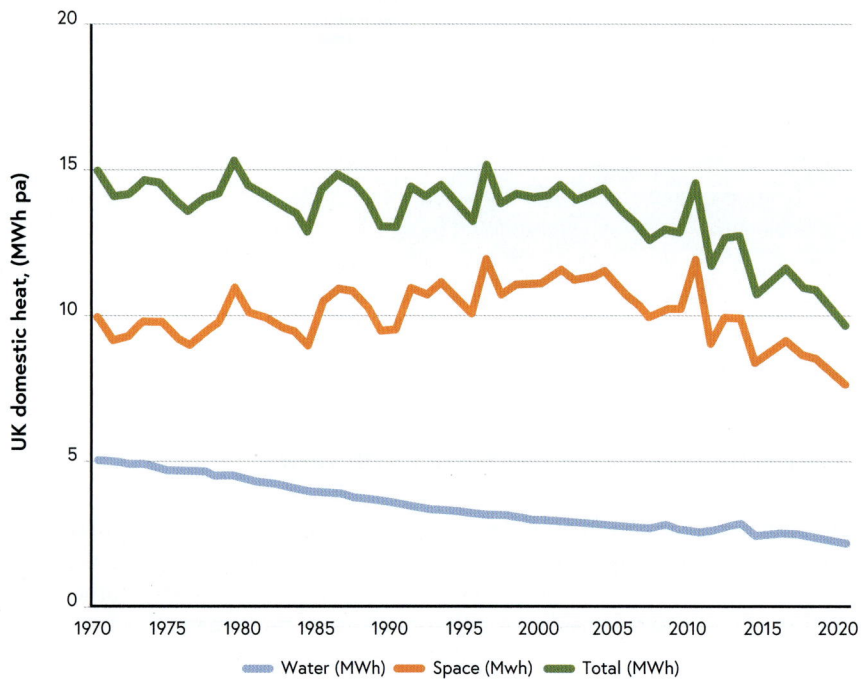

Space heating is predominantly affected by housing insulation. Figure 1.8 shows that after 2000 the UK started to see a sustained reduction in space heating demand. This was mostly driven by a growing realization that large cuts in CO_2 emissions needed to be made to address climate change and the UK's Kyoto Protocol target agreed in 1997[5]. The other major factor is weather, which can cause significant variations in demand. For example, 2010 was one of the coldest winters the UK had experienced in 30 years and this resulted in a 20 % increase in the annual demand for domestic space heat. Climate and weather also have a big impact on the daily variations in demand. Figure 1.9 shows the daily changes in gas demand, where a 40 % change in demand from one day to the next is not unusual during winter. Consequently, weather presents an operational challenge for the gas grid. However, the gas system has considerable storage capacity that includes underground storage as well as grid pipeline storage (referred to as line-pack). This, along with other operational tools, is valuable in ensuring gas demand is met (see www.nationalgrid.com/uk/gas-transmission/sites/gas/files/documents/Operational%20Overview%202018.pdf[6]).

---

[5]Eyre, NJ, and Mallaburn, PS *Lessons from energy efficiency policy and programmes in the UK from 1973 to 2013* Energy Efficiency 7 (1): 23–41. 2014

[6]An operational overview of the GB gas national transmission system can be accessed at: www.nationalgrid.com/uk/gas-transmission/sites/gas/files/documents/Operational%20Overview%202018.pdf

© The Institution of Engineering and Technology

Section 1 – Introduction

Figure 1.9   Daily profile of domestic gas demand

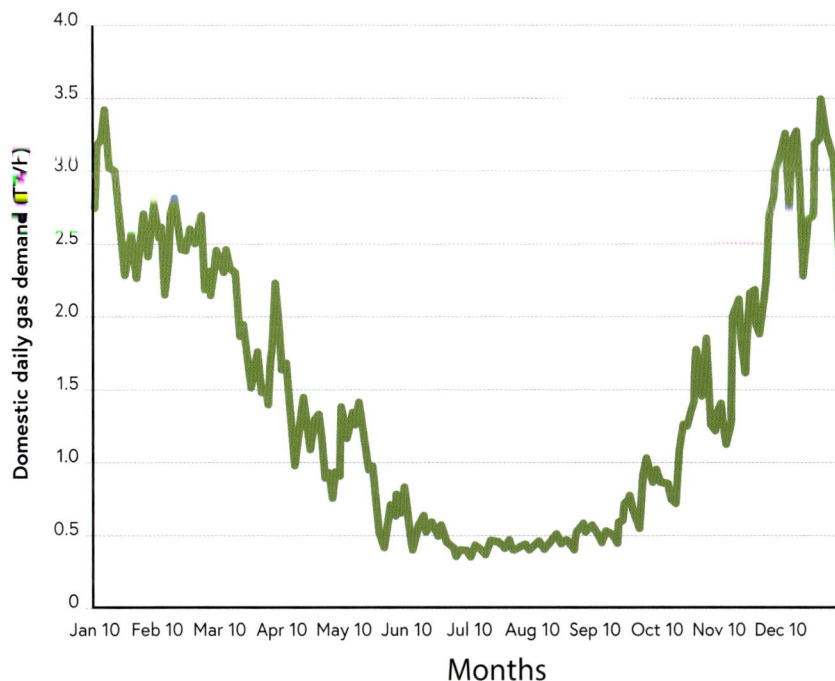

Figure 1.9 shows how gas consumption can vary across the seasons, with demand peaking in the winter months and dropping to its lowest levels in the summer months (attributed to hot water demands).

With the electrification of heat, balancing demand with generation will become an even greater challenge as the storage capacity is considerably less due to the much higher cost in, for example, batteries and pumped storage. But there are other options for balancing demand, such as demand-side management. This has been used for many years in industry where large consumers are asked to reduce demand at short notice. With the advent of smart controls in domestic households the potential of using demand-side management control of millions of appliances is seen as an important tool to balance demand with generation. However, this is, in part, dependent on consumer willingness to allow their appliances to be controlled by third parties.

1.7   How might the demand for heat change in the future?

All major transitions involve significant cost and changes to infrastructure. For the large-scale electrification of heat this will involve huge investment in power generation, transmission and distribution, including low-voltage connections to households. With the transition from natural gas, buildings will need to be upgraded to improve insulation levels, and changes to the buildings' space and hot water systems will also be required. If gas is used for cooking, then this will also need to be replaced. Projections of the scale of the electrification of heat range from a few million households to most of the existing households currently connected to the gas grid, i.e. 25 million households. This would mean transitioning up to one million households from natural gas to electricity every year over the next 25 years, or about 20,000 households every week. It is, therefore, important that projections of future demand for heat are made to underpin the planning required.

Section 1 – Introduction

To do this, it is necessary to identify the key factors that will influence demand. These include:

(a) household insulation improvements. The quality of the UK's housing stock is generally regarded as poor, particularly where insulation levels are concerned (see Figure 1.10). Over half of UK buildings were built before the Second World War and it is only since 2000 that significant energy efficiency improvements have been made. The government has indicated it wishes to see a 30 % improvement in energy efficiency and this is likely to have the most impact on the future demand for heat.

Figure 1.10 Typical insulation in a UK house

(b) household numbers. Population and household projections by the UK's four nations has resulted in an estimated increase in the number of households from 28 million to 32 million by 2040. On the assumption that these are new households requiring houses to be built to much higher standards of energy efficiency, the impact on heat demand is likely to be proportionately much less (around 5 %).

(c) climate change. Average UK winter temperatures are projected to increase by 2 °C. This would result in a reduction in space heat demand of around 20 %.

Combining these factors would result in a large reduction in space heat demand from around 250 TWh per year to around 150 TWh per year by 2050, with further reductions beyond this period. Hence the scale of the challenge is substantially reduced in terms of infrastructure, but also of the energy required at a household level. This illustrates the importance of projecting future demand for heat and the adherence to an ambitious and rigorous programme of building efficiency improvements.

Figure 1.11 projects the potential impact of climate change on heat demands up to 2100. Warmer winters are predicted to reduce demand for heat.

Section 1 – Introduction

Figure 1.11   UK domestic space heat projection illustrating the possible impact of climate change (Source https://es.catapult.org.uk/reports/domestic-heat-demand-study/ and the figures are based on Figures 1.7 and 1.16 of the report)

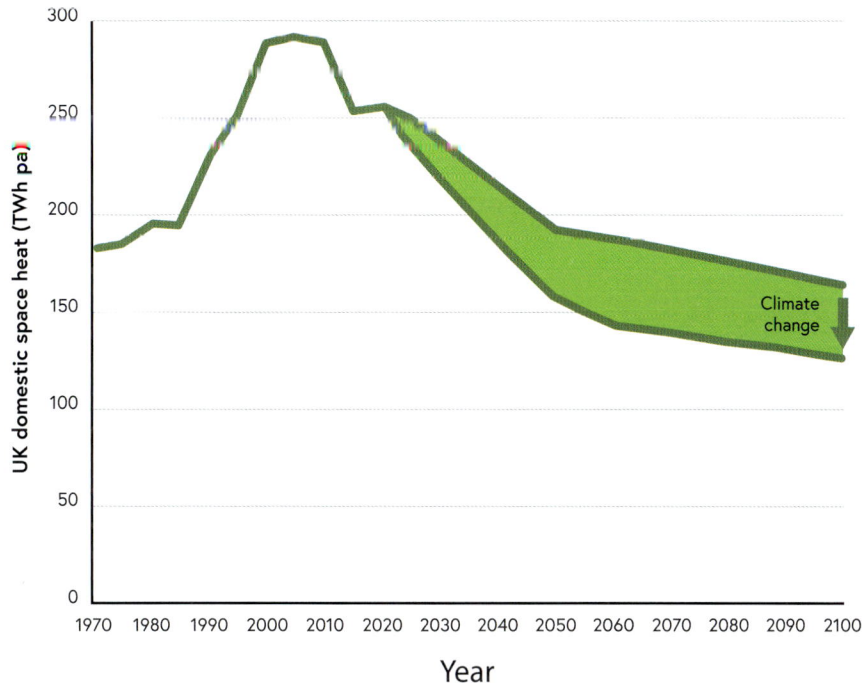

1.8   How can we decarbonize domestic heat?

To decarbonize heat, households need to transition from natural gas for heating to a low-carbon alternative. This transition is likely to be very disruptive, costly and is viewed as one of the biggest challenges to achieving net zero emissions. It is exacerbated by the poor quality of the UK's housing stock which has not only led to higher energy costs but has also meant that homes are more likely to be less comfortable and less healthy. Figure 1.12 illustrates the condition of the UK's housing stock. The vertical axis uses the measure heat loss parameter (HLP), which refers to the heat loss for each degree of internal and external temperature difference adjusted for floor area. Therefore, a house with a floor area of 100 m^2, an internal temperature of 20 °C, an external temperature of 0 °C and a HLP of 4 W/m^{2}K would have a heat loss of 8,000 W.

In Figure 1.12, the horizontal axis shows date of construction and the plots on the data curve show the number of households in millions. For example, there are 12 million households built before 1945 which have a HLP of 3.75 W/m^{2}K. Housing insulation improves after 1945 (HLP declines) but it is only after 2000 that a significant improvement starts to be seen. New builds are projected to have improved insulation levels, but the overall UK space heating demand impact will be small, unless substantial changes are made to the existing housing stock. For example, the weighted average HLP of all housing stock is 3.14 W/m^{2}K in 2015 but this only reduces to 2.78 W/m^{2}K, which is around 12 % (based on the assumptions made on future housing).

Figure 1.12 shows the historic and a projected (weighted) average HLP. It is heavily influenced by housing built before 1945 due to the number (12 million) and their poor insulation performance. The UK did not see any significant improvement in housing insulation until 2000.

Hence to achieve the 30 % improvement in housing insulation the focus needs to be on existing housing and particularly those houses built before 1945. By improving housing insulation, not only is energy consumption reduced but the capacity of the heating system can be reduced. This is much less important for gas-based heating systems as the size of the boiler is less sensitive to its cost, which is not

Section 1 – Introduction

the case for electric heating. Furthermore, a well-insulated and ventilated building provides higher comfort levels and contributes to a healthier living environment.

Figure 1.12 UK housing HLP 1945–2100 (Source https://es.catapult.org.uk/reports/domestic-heat-demand-study/)

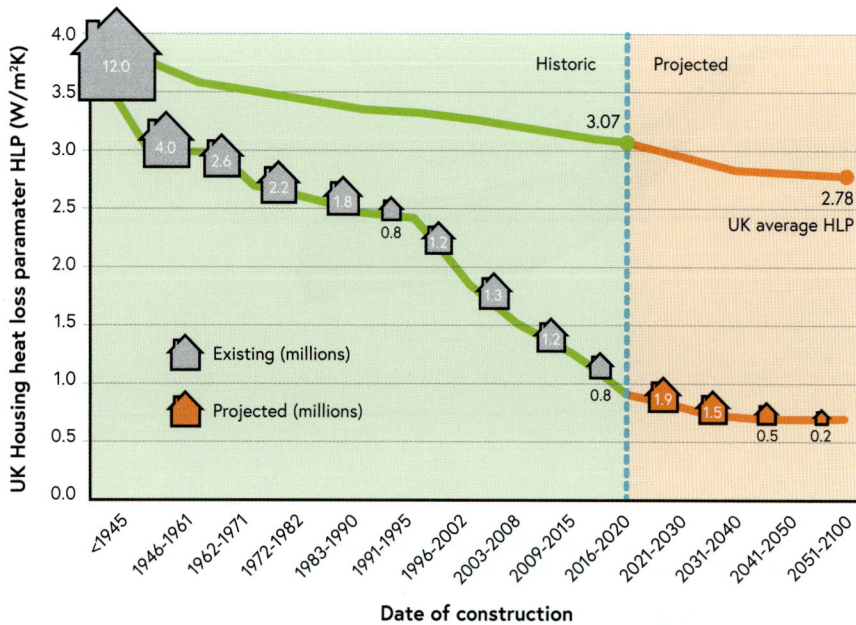

Figure 1.13 shows the impact of housing insulation levels on space heat demand. It assumes a typical space heating profile with operation in the morning and evening and with housing insulation based on HLPs shown in Figure 1.12.

Figure 1.13 Daily impact of improved insulation levels (Source https://es.catapult.org.uk/reports/domestic-heat-demand-study/)

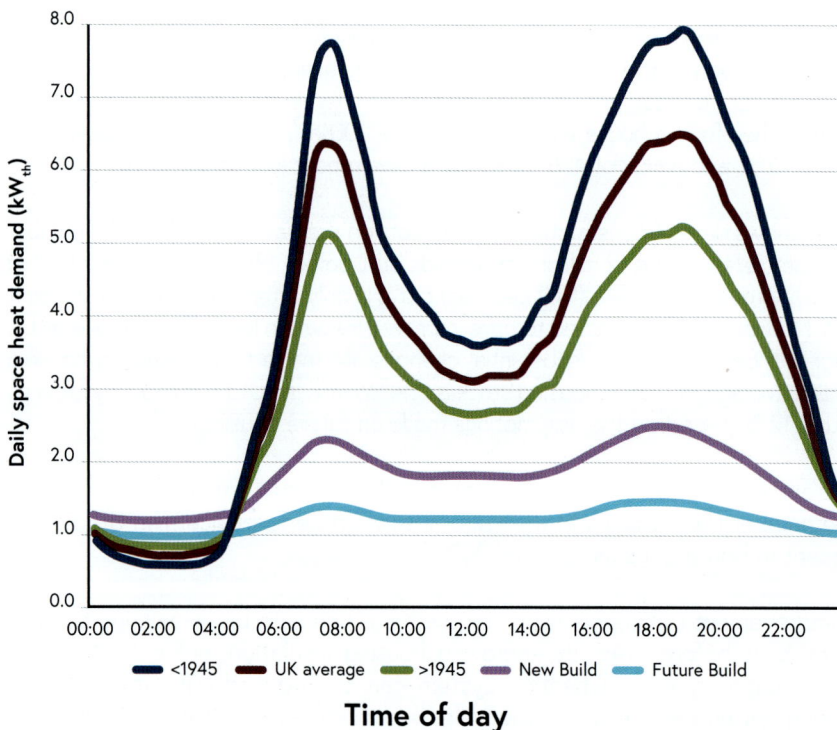

Buildings built before 1945 have the worst levels of housing insulation and therefore heat consumption is the highest, but also the heat demand profile has higher peaks to compensate for the higher levels of heat loss.

With improvements in housing insulation, space heat consumption is reduced but the daily heat demand profile is also less erratic. For new and future housing, the profile is much flatter and in addition to lower levels of heat consumption the peak demand is also much lower and therefore more suited to heat pump applications.

Another important feature is that the flatter profile is more suited to demand side control as it is far less sensitive to changes in space heating timer settings. This is useful to help manage network or grid demand, particularly when there are capacity constraints. With lower levels of housing insulation, the impact of changes to timer settings is more likely to result in lower comfort levels for householders due to deviations in internal temperature settings.

1.9 Low-carbon heating for large-scale deployment

There are several options for low-carbon heating, but those suitable for large-scale deployment may be categorized as:

(a) district/community heating;
(b) hydrogen heating with the existing natural gas network repurposed for use with hydrogen; and
(c) electric heating.

1.9.1 District/community heating

This is where heat for more than one building or area is produced at a central location and distributed through a network of insulated pipes. Heat sources can include:

(a) combined heat and power units. This is power generation with heat recovery, but to be low-carbon the power generation would need to be fuelled by a low-carbon source of energy, for example, hydrogen or natural gas with carbon capture and storage.
(b) network heat pumps. These can use a variety of heat sources, such as sea water, river water, lake water, ground and air.
(c) waste(d) heat recovery. This can be from an industrial process, such as buildings or a sewage system, and might also be combined with a heat pump to raise the temperature of the heat.
(d) solar thermal. This uses large solar thermal arrays to collect solar heat supplemented by heat storage.
(e) geothermal: this can be deep geothermal (up to 5 km) or shallow geothermal (around 100 m).

The main components of a district heating system are:

(a) heat storage: for additional capacity and backup;
(b) heat network: for connecting heat sources with buildings using supply and return pipes;
(c) pumps: for circulating the hot water;
(d) heat substations: for pressure and temperature changes;
(e) heat metering and control systems: for monitoring production and consumption; and
(f) heat exchangers/heat interface units: for connecting the heat network to the heating system in buildings.

1.9.2 Hydrogen heating

This is a colourless, odourless and non-poisonous gas that has an energy content around two and a half times that of natural gas (by weight) and a third of the energy density (by volume). When combusted the only by-products are water and nitrous oxide emissions, which will need to be controlled. Hydrogen gas boilers suitable for domestic households are under development and designed to permit operation on natural gas with subsequent conversion capability to hydrogen. The large-scale deployment of hydrogen would require the repurposing of the existing gas network from natural gas to hydrogen. Most of the existing low-pressure network has already been replaced with polyethylene pipes replacing iron pipes that are expected to be suitable for hydrogen, although investigations are ongoing. The transition to hydrogen would involve substantial investment in terms of production, storage and distribution.

There are several methods for producing hydrogen. These include:

(a) reforming. This is a chemical process whereby hydrogen is produced from methane (natural gas). However, the process produces large volumes of CO_2 which would need to be captured and stored for the hydrogen to be deemed low-carbon.

(b) electrolysis. This uses electricity produced by low-carbon energy sources to electrolyze water (split water into hydrogen and oxygen) to produce hydrogen.

(c) bioenergy with carbon capture and storage (BECCS). This uses a process to extract energy from organic materials. These materials can be from agriculture or forestry sources or from municipal waste. The process is considered low-carbon as the organic material would have absorbed carbon during its growth. Therefore, when it is processed it is only releasing the carbon previously absorbed as well as the stored energy. By adding carbon capture and storage the technology can provide negative emissions. This is regarded as extremely important to offset other hard to decarbonize sectors.

1.9.3 Electric heating

There are also several options for installing electric heat, which include:

(a) direct electric heating;
(b) electric storage heating; and
(c) electric heat pumps.

Direct electric heating

These include electric underfloor heating (UFH), infrared heaters, fan heaters, electric radiators and convection heaters. The capital/appliance cost is low but the running costs are comparatively high. They are normally used for supplementary heating and appliances can be portable. As infrared heaters use radiant heat, they are sometimes used for outdoor heating.

There is, however, a case for direct electric heating in well-insulated homes that have restricted space for external heat pumps. Direct electric heating is also an option in the provision of electric water heating where demand is low.

Electric storage heating

These have been used for many years in the UK, particularly in households not connected to the gas grid. By storing heat, they are mostly used with a two-rate tariff comprising day and night rates. Typically, the night rate would be approximately half the day rate, significantly reducing the running costs. Present day electric storage heaters are considerably more sophisticated with smart controls offering scope for flexible demand-side management.

Electric heat pumps

Heat pumps use a refrigerant cycle to transfer heat from a source such as air, ground or water to the medium to be heated, in this case the water in central heating systems. The cycle is similar to a home

Section 1 – Introduction

refrigerator where the source is within the appliance and the heat is transferred outside it. The cycle comprises of:

(a) **evaporation**: the refrigerant passes through a heat exchanger where it evaporates absorbing heat from the source;

(b) **compression:** the heated refrigerant is then compressed to a higher pressure using an electrically driven compressor;

(c) **condensation**: the compressed refrigerant passes through another heat exchanger located in the space to be heated and condenses, changing back to a liquid and releasing heat in the process, and

(d) **expansion**: the liquid refrigerant passes through an expansion device lowering its pressure prior to returning to the evaporator to repeat the cycle.

The source can be air for households (as shown in Figure 1.14), but it can also be ground and water if they are available. Heat pumps can be connected to a 'wet' heating system comprising of radiators. It may be possible to use the existing heating system when retrofitting, but radiators may need to be replaced with larger heat capacity units to compensate for the lower water flow temperatures. The main advantage of heat pumps is their operating efficiency, which typically averages 300 % thereby reducing running costs and the impact on energy production, storage, networks and carbon emissions compared with direct electric. However, although the electricity consumption of heat pumps is much less than other forms of electric heating, their capital cost is much higher.

Figure 1.14   Air source heat pump (ASHP) providing a wet heating system

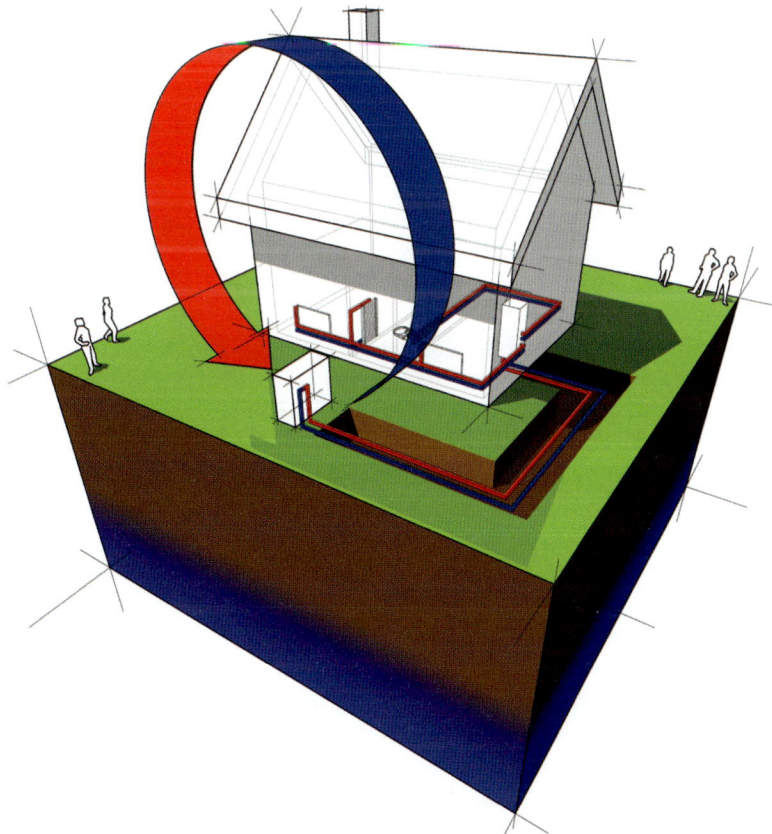

Heat pumps can also be used in an air-to-air arrangement where the heat source is air and the heat is transferred to air within the area to be heated. Typically, these are wall hung units, but an alternative

arrangement is combined with a building ventilation unit providing ducted heating. An advantage of air-to-air heat pumps is that they can be operated in reverse and provide cooling as well as heating.

A variation on the heat pump arrangement is the hybrid heat pump. This combines a heat pump with a gas boiler. Operation is mainly in heat pump mode with the gas boiler operating for 'peak' duty only. This could be when it is very cold and the heat pump is unable to provide sufficient heat, but it could also be to provide hot water on demand, similarly to a combi gas boiler. Hybrid operation substantially reduces the volume of gas consumed and the associated CO_2 emissions. Bivalent systems comprising the existing gas boiler operating in parallel with a new heat pump are also possible. In terms of cost, hybrid heat pumps might have a lower overall cost than a heat pump-only installation. This might be because a smaller heat pump may be suitable, the existing heating system can be used without radiator upgrades and hot water storage can be avoided. However, the boiler would need to operate on hydrogen – assuming that natural gas appliances will not be permitted in the future.

1.10   Economics

As households transitioned to natural gas in the 1970s and beyond, electric heating struggled to compete due to its high running costs. Despite the launch of a lower off-peak tariff (known as Economy 7) for use with storage heaters, electric heating was generally only competitive for households that were not connected to the gas grid where the only alternative was oil or solid fuel heating.

The price disparity between gas prices and electricity prices is still seen today. For example, when the Office of Gas and Electricity Markets (Ofgem) set the energy price cap in February 2021[7] they announced that, 'the equivalent per unit level of the price cap to the nearest pence for a typical customer paying by direct debit will be 19 p per kWh for electricity customers and 3 p per kWh for gas customers. More competitive and lower tariffs can be found but, in the worst case, it can cost more than six times as much to heat a home using electricity than gas. The economics are significantly helped by electric heat pump technology that generates three times as much heat for each unit of electricity consumed, but that still means the running cost is twice as high as gas. There is also the much higher capital cost of an ASHP installation to consider.

Figure 1.15 compares direct electric, gas and ASHP heating. The direct electric has the lowest capital cost of £2,000 per household and the ASHP has the highest cost at £10,000 per household, with gas heating in between at £5,000 per household. The running cost assumptions are £0.19/kWh for electricity and £0.03/kWh for gas but the ASHP is assumed to have an efficiency of 300 % whereas for direct electric and gas heating this is 100 % and 90 % respectively. This means that for annual heat demand of less than 2 MWh per year direct electric heating has the lowest cost, but above 2 MWh gas heating is the lowest. ASHP heating remains uncompetitive with gas heating but above 6 MWh per year its cost is lower than direct electric heating.

One of the factors that adversely effects electric heating arises from the addition of policy costs attributed to environmental and social costs. The CCC estimates that a householder switching from gas to electricity would incur an extra £100 per year due to these costs, despite reducing their carbon emissions in the process[8]. This is not enough to compensate for the higher electricity cost, but with the prospect of the UK government banning the sales of natural gas appliances for heating by the mid-2030s the competitive positioning of ASHPs, direct acting heating and electric storage will change and become more favourable.

---

[7] *The Sixth Carbon Budget - The UK's path to Net Zero* December 2020. Available at: www.theccc.org.uk/publications
[8] www.ofgem.gov.uk

Section 1 – Introduction

Figure 1.15   Economics of electrified heating

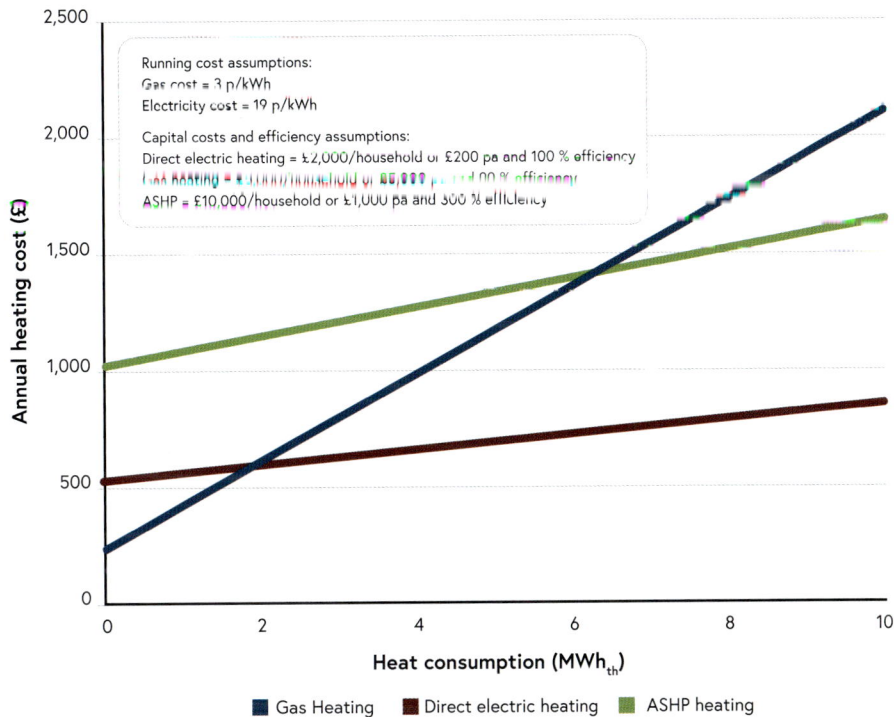

The chart shows annual heating cost (£) on the y-axis (0 to 2,500) versus heat consumption (MWh_th) on the x-axis (0 to 10).

Running cost assumptions:
Gas cost = 3 p/kWh
Electricity cost = 19 p/kWh

Capital costs and efficiency assumptions:
Direct electric heating = £2,000/household or £200 pa and 100 % efficiency
Gas heating = £3,000/household or £300 pa and 100 % efficiency
ASHP = £10,000/household or £1,000 pa and 300 % efficiency

Legend: ■ Gas Heating   ■ Direct electric heating   ■ ASHP heating

The business case for retrofitting heat pumps in existing housing stock will be more complex for homeowners, who will have to make the tough decision to either remain with their existing heating system or upgrade to heat pumps on environmental grounds.

1.11   Summary

The introduction of natural gas in the 1970s led to the widescale deployment of central heating and by 2015 central heating was being used by more than 85 % of UK households. However, it is a GHG and to meet the UK legislated net zero target by 2050, the UK will need to transition from natural gas to a low-carbon heating alternative.

A prerequisite of this transition is to improve building energy efficiency. The quality of the UK's housing stock is generally regarded as poor, and the government has indicated it wants to see a 30 % improvement in energy efficiency. Combined with the impact of warmer winters arising from climate change, improving building energy efficiency should result in a significant reduction in heat demand, more than offsetting the increase in household numbers as a result of population growth. The scale of the challenge to decarbonize heat is therefore substantially reduced.

There are several options for low-carbon heating and those suitable for large-scale deployment include different options in electric heating. It is anticipated that ASHP will have a major share of the future market for household heating. However, the economics are currently not favourable. This is due to a combination of high capital cost for the heat pump installation and building upgrades, particularly for older buildings with low levels of housing insulation, but also high running costs that include policy costs. With the prospect of the UK government banning the sales of natural gas appliances by the mid-2030s the competitive positioning of ASHPs should change and become more favourable.

© The Institution of Engineering and Technology

▆ Section 2

Fabric and comfort considerations

As the UK moves towards the use of more sustainable forms of energy, we must also treat it as a scarce and valuable resource regardless of how it is generated. The transition to electrified heat must include adequate consideration to the building fabric and thermal comfort conditions within a home. Typically, heat pump solutions operate at lower output temperatures and rely on the building fabric for better performance. It also makes sense to save every unit of heat from escaping the building so occupants are kept warm and valuable energy is conserved. However, building fabric upgrades are not just about energy saving as they also improve comfort, health, and wellbeing and can extend the usable space in many homes.

In this Section we will discuss passive heat and building fabric performance, including insulation levels and thermal comfort that are applied to electrified heat. The Section explores U-values, methods of improving insulation levels and sets out how to determine a building's heat loss to be able to select a suitably sized electrified heating solution.

2.1 Building orientation and solar gains

2.1.1 Introduction

Passive solar gains are a method of providing 'free heat' from the sun (see Figure 2.1), and therefore should be maximized wherever possible during heating seasons. Historically, little regard has been paid to the design of UK buildings in relation to solar orientation. However, as our understanding of building physics has progressed, we have become more aware of the benefits and challenges that the position of the sun presents when considering energy efficiency and thermal comfort.

Figure 2.1   A home being heated with passive heat

Section 2 – Fabric and comfort considerations

Along with the building fabric, passive heat is a key consideration for any active heating system, including electrified heat, by reducing reliance on and demand for heating.

The key considerations for electrified heat in any building, relating to operating factors and electricity costs, are:

(a) reduced demand and heat load; and
(b) lower flow and emitter temperatures than traditional fossil fuel-fired heating systems.

Key terms include:

(a) **load**: the instantaneous quantity of heat (or cooling) required to reach a specified set point (measured in W); and
(b) **demand**: a cumulative measure of heating energy expended over time (measured in Wh, kWh, MWh).

2.1.2 Building orientation and solar gain considerations

Considerations around solar gains vary greatly depending on the site climate and microclimate. These include:

(a) solar orientation: the path of the sun across the sky and the building's relative position; and
(b) exposure to wind: sites have a prevailing wind direction (i.e. the direction the wind blows most of the time).

In hot climates (known as 'cooling-led' climates), building orientation should be chosen to reject direct solar gain to reduce the likelihood of overheating, as well as its reliance on a cooling plant. Conversely, in a predominantly cold ('heating-led') climate, exposure to direct solar gain is beneficial in reducing reliance on an active heating system.

Difficulty arises in warm-temperate climates, such as the UK, where design conditions can range from very cold to very warm, and climate change projects extremes of temperature to become more exaggerated.

Theoretically, a building's orientation has a sizeable effect on a building's ability to comply with building regulations. The UK's Standard Assessment Procedure (SAP) is the methodology used to assess compliance with the Part L1A *Conservation of Fuel and Power in Domestic Buildings* of the Building Regulations. This compares a representation of an actual dwelling with a notional target, assessing both total regulated emissions – the dwelling emission rate (DER) – and the dwelling fabric energy efficiency (DFEE).

Dwelling orientation contributes significantly to both these metrics swaying a result by up to ±20 %, dependent on the extent and orientation of glazing area. In particular, this recognizes that south-facing glazing uses solar gains during the heating season, being net heat energy positive (that is, more heat is gained during the day than is lost at night), whilst glazing on north, east and west façades generally present a net loss of heat through the heating season.

2.1.3 Occupancy zones in buildings

Traditionally, building spaces are arranged according to their function, with daytime living spaces distributed around the ground floor and bedrooms and bathrooms occupying upper storeys.

Section 2 – Fabric and comfort considerations

The arrangement of spaces within buildings is vitally important in achieving an effective energy balance, so that heating demand and load can be minimized. There are four considerations relating to occupancy zones:

1. **orientation:** south-facing zones (those with south-facing openings) receive more useful solar gain;
2. **heat demand:** the quantity of heat required to reach the set point temperature and maintain it throughout the occupancy period;
3. **the occupancy profile:** the times throughout the day and night that a space is occupied; and
4. **internal thermal gains:** heat generated within the space by lighting, occupancy and equipment.

2.1.4 Sun path and solar gains

Being in the Northern Hemisphere, the UK experiences a southerly solar track. This means that between sunrise and sunset, the sun moves across the south in an east to west direction.

For building orientation, this presents both opportunities and challenges. For passive solar heat gain, which is useful during the heating season, solar radiation will be received mainly through openings in the south façade, amounting to a net gain of heat and reducing demand from an active heating system. Daylight hours are relatively short during winter months and the solar altitude is far lower in winter; in the UK, the peak solar altitude at the winter solstice is around 10° from the horizon.

In summer months, the sun follows the same southerly path but with longer daylight hours. With more intense solar radiation during these months, there is a risk of overheating in buildings oriented with large areas of south-facing glazing. It is therefore important to consider shading measures to mitigate any risk of overheating. At the summer solstice, the peak solar altitude is around 64° from the horizon.

The solar azimuth (its position relative to true north) is also a significant consideration in determining optimum building orientation.

At the winter solstice, the sunrise occurs at 130° (east) and sets at 230° (west). This means that the sun's path and exposure to it will be predominantly across the south façade of a building, with east- and west-facing windows capturing little solar gain.

At the summer solstice, the sunrise occurs at 50° (east) and sets at 310° (west). In this case, east, south and west façades receive significant solar irradiance, but at a time of year when there is little or no demand for space heating.

2.1.5 Passive solar heating

Passive solar heating refers to the ability of a building to make use of passive solar gains as shown in Figure 2.2.

Effective passive solar heating is dependent on:

(a) building orientation and massing;
(b) glazing area and orientation; and
(c) glazing solar properties.

Orientation is an important design factor to consider for passive heating. It is normal practice to plan the layout of buildings and internal rooms to maximize solar gain, but also to avoid overheating issues.

Section 2 – Fabric and comfort considerations

Figure 2.2   Passive solar heating and the creation of heat zones in a home

The following should be considered for each façade:

(a) south façade:
  i.   allow for predominantly south-facing glazing to capture useful solar gain;
  ii. ensure proper shading provision that permits low-level solar penetration in winter, but protects against it in summer;
  iii. employ thermal modelling services to identify potential risk of overheating; and
  iv. consider the use of solar control glazing where necessary;
(b) east and west façades:
  i.   allow for some glazing area, but less than on the south façade; and
  ii. employ shading where necessary, particularly on west façades to minimize the risk of afternoon overheating and low-level glare;
(c) north façade:
  i.   minimize north-facing glazing to prevent excessive heat loss.

2.1.6 Thermal gain zones

Understanding that, in the Northern Hemisphere, useful heat is received through openings by spaces with a south-facing aspect, there are certain considerations that can be made to lower the heating demand from an active heating system.

To reduce the heating load, it is recommended that:

(a) spaces with higher demand (such as living rooms or dining rooms), are situated on the south façade.
(b) spaces with lower demand, or ones which are sporadically occupied (such as bathrooms, utility spaces and storage areas) are situated on the north façade.
(c) bedrooms are situated on the north façade, since they are typically only occupied at night when thermal comfort can be regulated with clothing.
(d) separate control zones are set up so that heat can be distributed to spaces when required, but not when the set-point temperature has been reached. This would typically mean having separate thermostats in north- and south-facing zones so that each space is individually controllable.

Section 2 – Fabric and comfort considerations

These recommendations are particularly effective when implemented in conjunction with mechanical ventilation and heat recovery, so that warm air can be distributed around the dwelling, further reducing reliance on an active heating system.

2.1.7   Solar gain management

A building designed to effectively capture useful solar gains can also be prone to risk of overheating. Management of solar gains is therefore crucial to mitigating this risk and numerous measures can be taken to block unwanted solar radiation. These include g-values (otherwise known as solar heat gain coefficients, window solar factors or total energy transmittance).

The solar energy transmittance coefficient given to glazing that quantifies the proportion of solar radiation able to pass through a glazing unit relative to that received by its surface can be determined by the following equation:

$$\text{g-value} = \frac{\text{total solar heat gain}}{\text{incident solar radiation}}$$

Higher g-values allow for the passage of a greater amount of solar energy, whereas lower ones limit the passage of solar energy. Typical g-values for standard glazing are shown in Table 2.1.

Table 2.1   Glazing g-values

Type of glazing	g-value
Single-glazed float glass	0.9–1.0
Double-glazed glass	0.4–0.7
Triple-glazed glass	0.25–0.6

Note that solar energy transmittance does not refer to visible light transmittance. Purpose made 'solar control' glass limits heat gain whilst permitting the passage of visible light. This type of glass can be considerably more expensive, so it is often more cost-effective to employ shading devices where possible. Solar control films can also be used to lower solar energy transmittance, but these often also reduce visible light.

External shading

External shading devices, such as *brise soleil* shown in Figure 2.3, or external blinds prevent the ingress of unwanted solar radiation. Devices can be designed using geometric principles with knowledge of solar azimuth and altitude at a particular time of year so that useful solar gain can be permitted during the heating season whilst also providing protection against high level intense gain. This can be achieved with either fixed shading devices or removable/retractable ones, such as externally mounted awnings or roller blinds.

Figure 2.3   South-facing *brise soleil* solar shading

Note: This example also shows external blinds on the windows to provide additional shade

Natural shading

Sometimes the simplest solutions are the best, and topographical or floral barriers can offer sufficient shading to prevent overheating. Deciduous trees and bushes naturally adapt their shading coefficients with the seasons: they offer shade in summer before dropping their leaves in winter and allowing useful solar gain.

Internal shading

Solar radiation passes through transparent or translucent elements as shortwave radiation and is converted to longwave radiation as it warms the glass and is emitted inside the space. Longwave radiation cannot pass back through transparent materials, resulting in a greenhouse effect. Internal shading, such as blinds or curtains, whilst cost-effective and easily adjustable, do not offer robust protection against the risk of overheating, as solar energy has not been prevented from entering the building space. Internal blinds however can offer finely adjustable protection against glare, so are useful in winter.

Interstitial shading

A relatively recent innovation is the installation of blinds between glazing panes. These are often more effective at blocking solar energy than internal blinds, however they are often an expensive solution and also render glazing units less effective at retaining heat since they cannot be fully sealed.

The recommended strategy for solar gain management is first to identify areas at risk of overheating using thermal modelling or geometrical calculation. Once identified, the most appropriate shading option can be employed. External shading is the most effective option in protecting against excessive solar gain and although it is used widely in most European countries, its uptake in the UK is limited.

2.2   Fabric thermal properties and air leakage

2.2.1   Building fabric performance

The building envelope refers to a building's thermal enclosure and its fabric refers to the combination and composition of the materials used to construct and isolate it from external conditions (see Figure 2.4).

Section 2 – Fabric and comfort considerations

Figure 2.4   Cavity wall insulation being installed on a new build home

The transmission of heat occurs through three main mechanisms:

1. **conduction.** The diffusion of internal heat within a static (rather than fluid) body resulting from a temperature difference across it. This concerns the physical materials used in the fabric and their resistance to conductive heat transfer, which is addressed by using materials with high thermal resistance/low thermal conductivity.
2. **convection.** The movement of heat through a fluid (such as air). This occurs in buildings via forced or free ventilation. This includes infiltration (unregulated air movement through cracks and gaps in the building fabric), which is addressed through airtightness or air permeability.
3. **radiation.** The process by which heat is emitted directly via electromagnetic waves to a point of absorption without an intermediary medium. This occurs where heat conducted through a material is emitted from its surface, which is addressed using low emissivity materials and coatings.

A thermally robust building envelope is the key to an economical and effective electrified heating system. A building envelope that addresses each of these heat transfer mechanisms adequately will retain heat effectively.

2.2.2   U-values

U-values are thermal transmittance coefficients for building elements and are frequently quoted in reference to elements' abilities to retain heat. A low U-value indicates high resistance to the passage of heat (and, in turn, low conductance).

U-values are calculated from the summed combination of thermal resistances of each material in combination with their thicknesses (in metres), together with internal and external surface resistances and measured in W/m^2K.

There are three key terms relating to U-value calculations (since the requisite thermal resistances are not always given in materials specifications).

Section 2 – Fabric and comfort considerations

Thermal resistance (R-value)

Thermal resistance is a heat property and a measurement of a temperature difference by which an object or material resists a heat flow. Thermal resistance is calculated using the thermal conductivities of a material together with its thickness. This is calculated as:

$$R = \frac{l}{\lambda}$$

where:

l is the thickness of material (in metres)

λ is the thermal conductivity

Thermal conductivity (λ)

Thermal conductivity is a value normally given in a material specification and refers to a material's ability to transmit heat by conduction. This is expressed as a lambda value (λ), and is measured b W/mK. Materials with high conductivity are those that transmit heat most easily, such as metals, whilst those with low conductivities, such as plastics, mineral fibres, air and inert gases, make the best insulation materials.

The most effective insulation materials are those that not only have low conductivities, but that also trap or suspend inert gases or air. Popular insulation materials are therefore aerated with these gases, enabling robust layers in building elements highly resistant to the passage of heat.

This is the reciprocal to thermal resistivity:

$$r = \frac{1}{\lambda}$$

Emissivity

A further property related to a material's ability to retain or transmit heat is emissivity. The emissivity of the surface of a material is its effectiveness in emitting energy as thermal radiation, hence materials with low emissivity coatings are more resistant to emitting heat by radiation.

Typically, smooth, hard materials (such as glass) have high emissivity properties whereas those with rough surfaces (such as most insulating products) have a lower emissivity.

2.2.3 Building Regulations and limiting thermal parameters

Approved Document Part L *Conservation of fuel and power* of the UK Building Regulations (for England and Wales) is split into four different sections to differentiate between newly constructed and existing domestic and non-domestic buildings:

1. L1A: new dwellings;
2. L1B: existing dwellings;
3. L2A: new non-domestic buildings; and
4. L2B: existing non-domestic buildings.

Section 2 – Fabric and comfort considerations

Each of these set minimum energy and carbon emissions standards for new buildings and renovation works carried out to existing buildings with reference to the fixed building services of heating, cooling, lighting, ventilation and hot water generation.

Crucially, the regulations also set minimum standards for building fabric and external envelope, by way of upper limits on U-values and air permeability. For domestic buildings, Part L1A sets the maximum limiting values for new domestic buildings (but note that limiting values for non-domestic buildings are different) Table 2.2 lists those limiting values along with those under consultation in Part L1A 2021.

Table 2.2   Limiting fabric values for thermal elements (Reproduced from Approved Document Part L1A: 2013 edition with 2016 amendments for use in England, and adapted to include values from Part L1A 2021 consultation)

Limiting fabric parameters	Current (L1A 2013)	Proposed (L1A 2021)
Roof	0.20 W/m^{2}K	0.16 W/m^{2}K
Wall	0.30 W/m^{2}K	0.26 W/m^{2}K
Floor	0.25 W/m^{2}K	0.18 W/m^{2}K
Party wall	0.20 W/m^{2}K	0.20 W/m^{2}K
Swimming pool basin	0.25 W/m^{2}K	0.25 W/m^{2}K
Windows	2.00 W/m^{2}K	1.60 W/m^{2}K
Roof windows, glazed roof-lights	2.00 W/m^{2}K	2.20 W/m^{2}K
Pedestrian doors	2.00 W/m^{2}K	1.60 W/m^{2}K
Air permeability	10.00 m^3/(h.m^2) at 50 Pa.	8.00 m^3/(h.m^2) at 50 Pa. 1.57 m^3/(h.m^2) at 4 Pa.

It is recognized that these values will often need to be significantly improved upon to pass the DER and DFEE indicators, as calculated by the SAP.

Updates to regulations occur periodically and checks should be carried out to ensure that the correct regulatory values are used. Notably, the *Future Homes Standard* (currently under consultation and proposed to take effect in 2025) suggests:

(a) proposals to introduce a new overheating mitigation requirement in the Building Regulations for new homes;
(b) improvement in standards when work is carried out in existing homes;
(c) reconsultation on the *Fabric Energy Efficiency Standard*, as well as other standards for building services in new homes and guidance on the calibration of devices for airtightness testing; and
(d) changes to Part F (ventilation) of the Building Regulations and its associated Approved Document guidance.

For existing domestic buildings (and extensions to them), Part L1B sets minimum standards for new and existing thermal elements such as walls, roofs and floors; controlled fittings such as windows and doors and fixed building services such as heating, lighting and ventilation.

Requirements for Part L1B depend on the extent of renovation and fittings and services being replaced, with the aim that any replacements should improve the energy efficiency of the building. Minimum standards for thermal elements and controlled fittings are shown in Table 2.3.

Section 2 – Fabric and comfort considerations

Table 2.3 Limiting fabric values for new and retained thermal elements

Thermal element/controlled fitting	New U-value (W/m²K)	Retained (upgraded) U-value (W/m²K)
Roof	0.16–0.18	0.16–0.18
Wall	0.28	0.3–0.55
Floor	0.22	0.25
Windows/doors	1.6–1.8	n/a

The prevailing version of Part L1B should be consulted for exact details and to establish the minimum standard that improvements to a building's energy efficiency must attain. Depending on the extent of works, a large dwelling may be subject to consequential improvements, whereby renovating a large proportion of the fabric will trigger the requirement to improve the rest of the dwelling in line with minimum standards.

Any improvements made to fixed building services, such as heating, ventilation or lighting appliances must be made in line with the minimum standards set out in the *Domestic Building Services Compliance Guide*.[9]

2.2.4 Airtightness and testing

Air leakage can account for a significant proportion of heat loss in a poorly sealed building. Whilst good fabric attributes are vital to lowering heat demand, it is recognized that good airtightness is often more effective in preventing heat loss than insulation, particularly in existing buildings where providing a continuous thermal line is more challenging.

Airtightness is a term used for the resistance of the building envelope to infiltrations with ventilators closed and is part of The Building Regulations Part L minimum performance standards for both domestic and non-domestic buildings. The maximum permitted air permeability rate is $10\,m^3$ of air per square metre of building fabric per hour, using a test pressure of 50 pascals (m^3/m^2.h@50 Pa). For good practice however, air permeability results should be significantly lower than this. For domestic buildings, average performance is around $5.0\,m^3/m^2$.h@50 Pa, whereas good performance would be considered $3.0\,m^3/m^2$.h @50Pa, or lower.

Wherever infiltration occurs, there is a corresponding exfiltration elsewhere in the building. During summer, infiltration can bring humid outdoor air in, whilst in winter, exfiltration can result in moist indoor air moving into cold air cavities. This can result in condensation and ultimately mould growth and rot.

Air permeability is tested using an air pressure test. Traditionally, a calibrated fan or compressed air vessel is used to gradually pressurize (or depressurize) a building with all openings closed to a specified pressure, recording the pressure differential at each step. In typical testing applications, the total air flow required to achieve a pressure differential of 50 Pa is calculated and divided by the total building envelope area. This provides the leakage rate in m^3/m^2.hr@50Pa, or in other lower static pressure testing methods is compared against a specified lower pressure. A low air permeability indicates high resistance of a sealed building to heat loss by convection.

---

[9] *Domestic Building Services Compliance Guide* UK Government, London, 2013. Available at: https://assets.publishing.service.gov.uk/government/uploads/system/uploads/attachment_data/file/697525/DBSCG_secure.pdf

Section 2 – Fabric and comfort considerations

Air testing can also be useful in existing buildings to establish where energy efficiency can be improved. If a building has a low air permeability, it will be more economical to focus improvements on insulating the fabric. If the air permeability is high, efforts to better seal the building are recommended before upgrading fabric elements.

2.2.5 Improvements to existing fabric

A major challenge for the integration of electrified heat into existing buildings is the prerequisite measure of upgrading the building fabric (see Figure 2.5) so that heat generated can be retained to ensure a resilient, economical and efficient electrified heating system. This is particularly important where electric resistance heating is proposed to prevent high running costs, and that would otherwise render it an unviable alternative to conventional fossil fuelled alternatives.

Figure 2.5   External wall insulation being retrofitted

Improvements to thermal performance (both thermal transmittance and airtightness) are worthwhile investments ahead of electrified heat integration, but there are considerations around cost and internal space that can obstruct its implementation.

Airtightness

Airtightness should be addressed first as it can significantly reduce ventilation heat loss. Ensuring windows and doors are properly sealed (or replaced to modern standards), caulking cracks or re-plastering walls with barrier membranes, installing shutters on ventilators and sealing redundant penetrations (such as gas pipes or flues if no longer used) are relatively inexpensive ways to improve airtightness.

Section 2 – Fabric and comfort considerations

Thermal transmittance

(a) Roofs. If a dwelling has a cold roof space (i.e. it is insulated at ceiling level), simply maximizing insulation depth is a positive first step. In the case of warm roofs (i.e. they are insulated at the pitch), measures may be taken to replace insulation with lower conductivity materials (such as phenolic foam or expanded polystyrene boards). This is likely to involve pulling down and replacing ceilings. It should be noted, however, that roof spaces are required to have adequate ventilation in to prevent condensation.

(b) Windows and doors. These are seen to be 'easy wins', since modern doors and windows are far superior to older ones in both reducing thermal transmittance and infiltration. Replacement of single with double glazing will significantly reduce heat loss, as will the replacement of older double glazing with modern equivalents, particularly where metal frames are replaced with plastic, softwood or composites. Where glazing cannot be replaced due to conservation concerns, well-sealed secondary glazing can be effective in reducing thermal transmittance.

(c) Floors. Since most properties in the UK are built directly onto the ground with no 'crawl space', insulating them must be done internally. The main issue with this is the loss of internal floor-ceiling height, so very low conductivity materials must be used which are also resistant to compression. Board insulation, such as polyurethane or polyisocyanurate are materials typically used for this application, since their low conductivity means thinner layers can be used. For suspended floors, where it may be possible to lift floorboards, insulation such as mineral wool can be installed and re-covered. Replacing the floorboards with interlocking chipboard or similar can also greatly improve airtightness. Where underfloor heating is proposed, floors must be insulated to prevent significant heat loss.

(d) Walls:

    i. **external**. If walls are to be insulated, it is often preferable to insulate externally, since it is more likely to provide a continuous insulation line, improve airtightness and will not impact the internal floor space. This typically involves installing a vapour barrier, bonding insulation boards or batts to the wall and then rendering or cladding. However, this is often more expensive due to working at height access requirements.

    ii. **cavities**. Cavity wall construction has been the typical method of construction in the UK throughout the twentieth century and twenty-first century, with the principal benefit of providing a physical break to damp penetration from the outside in. This is also an opportunity to provide a continuous thermal line, with building regulations making this virtually compulsory in new buildings from the late twentieth century. However, doing this retrospectively must be done with care and consideration to preventing the passage of moisture across the cavity. The installation of porous materials must be avoided; retrofitted cavity wall insulation originally used materials such as recycled newspaper, which would saturate with water and track moisture across the cavity, leading to damp issues (particularly if installed without a vapour or moisture barrier). More recently, installers have favoured the use of expanded polystyrene beads, rock wool, mineral wool or phenolic spray foam that are more resistant to the passage of moisture.

    iii. **internal**. A major drawback of internal insulation is the loss of internal floor space. Achieving a meaningful reduction on thermal transmittance, particularly on uninsulated solid walls, usually means installing 50 mm or more of low conductivity board insulation, which can subtract notable floor space depending on the dwelling size. It is also more difficult to provide a continuous thermal line due to the presence of internal wall junctions. Where internal insulation must be used, breather membranes should be used to reduce the risk of interstitial condensation.

2.2.6   Ventilation

Ventilation is the key component in ensuring adequate indoor air quality, since there are many aspects of indoor living that contribute harmful pollutants into the air. As buildings have become more airtight to retain heat, it has been necessary to ensure ventilation for an adequate supply of fresh air and to remove moisture from internal air. Approved Document Part F of the Building Regulations dispenses many conditions around ventilation and specifies minimum ventilation requirements for domestic and non-domestic applications.

Section 2 – Fabric and comfort considerations

The requirements for new dwellings are shown in Table 2.4

Table 2.4   Approved Document Part F: 2010 edition with 2013 amendments, for use in England

Extract ventilation rates			
Room	**Intermittent extract**	**Continuous extract**	
	Minimum rate	**Minimum high rate**	**Minimum low rate**
Kitchen	30 l/s adjacent to hob; or 60 l/s elsewhere	13 l/s	Total extract rate should be at least the whole dwelling ventilation rate given below.
Utility room	30 l/s	8 l/s	
Bathroom	15 l/s	8 l/s	
Sanitary accommodation	6 l/s	6 l/s	
Whole dwelling ventilation rates			
	Number of bedrooms in dwelling				
	1	**2**	**3**	**4**	**5**
Whole dwelling ventilation rate (l/s)	13	17	21	25	29

Notes: a) In addition, the minimum ventilation rate should not be less than 0.3 l/s per m^2 of internal floor area. (This includes floors, e.g. for a two-storey building add the ground and first floor areas).
b) This is based on two occupants in the main bedroom and a single occupant in all other bedrooms. This should be used as the default value. If a greater level of occupancy is expected add 4 l/s per occupant.

Methods of ventilation can vary depending on the age, situation, designed occupancy and thermal performance of a building. In older buildings with high air leakage, it is possible that infiltration alone provides adequate fresh air. In newer buildings which are more airtight, purpose-built ventilation solutions must be employed by natural means, such as window openings, passive stacks and trickle vents or forced convection via intermittent or constant extract, or whole-house mechanical ventilation.

Actual ventilation provision must be measured as part of the commissioning process for mechanical ventilation systems. Whilst manufacturer's flow rates enable the correct specification of ventilation products, it is the duty of the commissioning process to ensure sufficient rates are delivered and to avoid over-ventilation, which would result in excessive heat loss.

Of course, any ventilation method that exchanges cold external air with heated internal air is a mechanism for heat loss. An optimal solution to this is to install heat exchangers between the incoming and outgoing air streams so that heat can be 'recovered' into the fresh intake air. Mechanical ventilation heat recovery (MVHR) presents an advantage for mechanical ventilation during winter, since there is no opportunity to recover heat using natural means (for example, openable windows).

2.2.7   Passivhaus

There are a number of voluntary standards available that certify buildings in achieving a particular aspect of performance. These include LEED (Leaders in Energy and Environmental Design) and BREEAM (Building Research Establishment Environmental Assessment Method), which are holistic environmental rating

Section 2 – Fabric and comfort considerations

methods, WELL and FitWell, which principally assess health and wellbeing aspects in non-domestic buildings, and Passivhaus, which assesses both indoor environmental quality and exceptional energy performance. Passivhaus (otherwise known as Passive House, see Figure 2.6) is a building standard developed in Germany and is known to be among the most robust and rigorous in the world. It focuses heavily on occupant comfort and indoor environmental quality, with substantial energy savings due to the stringent assessment measures. These buildings tend to be highly suitable for most forms of electrified heat.

Figure 2.6 A Passivhaus showing external shading and low-angle glazing for passive heating

Whilst developed primarily as a design standard for domestic buildings, Passivhaus can be applied to most building types, with standards for both new build and existing buildings. Passivhaus buildings consider virtually all the aspects discussed in this Section, focussing on the minimization of heating demand and load.

A building can only be certified as a Passivhaus if it complies with a rigorous design and assessment, and the six key criteria are shown in Table 2.5.

Table 2.5 Passivhaus project criteria

Airtightness	0.6 air changes per hour (ACH) @ 50 Pa (n50)
Surface temperature (windows)	>17 °C
Summer overheating	Maximum 10 % > 25 °C as determined by the Passivhaus Planning Package (PHPP)
Ventilation	~30 m^3/h/person
Heating	15 kWh/m^2/yr @ 20 °C or 10 W/m^2 heating load
Primary energy	135 kWh/m^2/yr

Airtightness

The maximum airtightness allowance of 0.6 ACH, which differs from the air permeability used for building regulations compliance, is determined using the average of both positive and negative pressure tests. Achieving this stringent target is the shared responsibility of designers and contractors.

Section 2 – Fabric and comfort considerations

Surface temperature

A minimum internal surface temperature of 17 °C when the space is heated to 20 °C is specified to promote thermal comfort and minimize risk of condensation formation. Radiant asymmetry discomfort occurs when a body experiences a significant temperature difference on one side (for example, a cold floor or window) from the ambient temperature on the other. Passivhaus buildings address this using high levels of continuous insulation, minimization of thermal bridging at junctions and almost always triple-glazed windows.

Summertime overheating. A maximum limit of 10 % annual hours exceeding 25 °C is imposed, as assessed by a proprietary tool known as the PHPP or other appropriate measure. This is to ensure that highly insulated and sealed buildings can ventilate excessive heat gains and reduce the risk of overheating.

Ventilation. Due to the highly airtight building envelope, a reliable ventilation strategy is necessary to provide sufficient fresh air for occupants (30 m^3/h/person), keeping pollutant levels down. Passivhaus buildings typically use full MVHR to provide fresh air and recover as much heat as possible through a heat exchanger. Stale air is extracted from utility zones (such as bathrooms and kitchens) and supplied via separate air stream plate heat exchangers into living spaces, which distribute warm air throughout the building. Whilst this uses energy to drive fans, the benefits of recovering and maintaining heat mean that annual heating demand is kept to a minimum. The marked reduction in heating energy whilst maintaining a reliable fresh air supply makes this approach superior to natural ventilation.

Heating. The extremely low heating demand limit of 15 kWh/m^2/year, or alternatively a maximum heat load of 10 W/m^2 at a set point temperature of 20 °C, is key to ensuring very low energy operation. For this reason, a typical Passivhaus dwelling using central air distribution can require as little as a single heat emitter and makes them highly suitable for many electrified heat solutions. By contrast, standing heat loss from a standard new build house compliant with the minimum standards in Part L1A of the Building Regulations could be 25–50 W/m^2, whereas an uninsulated existing dwelling could exceed 100 W/m^2.

Primary energy. In simple terms, this is a measure of the embodied energy intensity of the fuel mix within the Passivhaus building. Each fuel (for example, grid electricity or mains gas) is given a determined primary energy factor and this is used to determine the building's effective primary energy intensity for each year of operation. Grid electricity has the highest energy intensity at over 3 kWh/kWh supplied, meaning that both space and water heating using electricity (as well as other processes such as lighting and auxiliary loads) must be as efficient as possible to remain within the requisite 135 kWh/m^2/year. It is therefore often necessary to specify heat pumps in Passivhaus buildings for both space and water heating due to their superior efficiency, which can surpass the primary energy requirements.

Each of these criteria are assessed using a PHPP. This is a spreadsheet-based utility that is populated with every detail of the building's location, situation, form, fabric, services, fuel mix and renewable energy systems. This can be used as a tool from the project outset, informing all aspects of a building's design to achieve the criteria listed in Table 2.5.

2.3 Thermal comfort

Thermal comfort is a subjective concept which can be theorized, but never definitively defined, since all people experience their surrounding environment differently – what may constitute comfortable conditions for one person may not for another. Theories around thermal comfort therefore centre around predicting the degree to which a person will find their thermal environment satisfactory and the proportion of a population that may find it comfortable.

Section 2 – Fabric and comfort considerations

2.3.1 Defining thermal comfort

BS EN ISO 7730:2005 *Ergonomics of the thermal environment. Analytical determination and interpretation of thermal comfort using calculation of the PMV and PPD indices and local thermal comfort criteria* defines thermal comfort as 'that condition of mind which expresses satisfaction with the thermal environment', which is when a person is feeling neither too hot nor too cold. Similarly to buildings, achieving thermal comfort is a heat transfer energy balance.

There are two main indices used to quantify thermal comfort.

1. **Predicted mean vote (PMV)**

   Given a set of environmental parameters, the PMV index uses a seven-point scale to assess the expected level of satisfaction using a cross-section of building occupants. The scale is set as:

   +3 = hot
   +2 = warm
   +1 = slightly warm
    0 = neutral
   −1 = slightly cool
   −2 = cool
   −3 = cold

2. **Predicted percentage dissatisfied (PPD):**

   This is 'an index that establishes a quantitative prediction of the percentage of thermally dissatisfied people who feel too cool or too warm'. For the purposes of BS EN ISO 7730, thermally dissatisfied people are 'those who will feel hot, warm, cool or cold', as per the PMV model. The acceptable percentage varies depending on the activity, but for most purposes, it is generally accepted that a thermal environment is deemed satisfactory if the PPD index score is less than 10 %.

These calculations can be performed manually, but it is now more common to use specialist software, such as Dynamic Simulation Modelling.

2.3.2 Environmental comfort factors

The main components making up thermal comfort have been identified as:

(a) **air temperature**. The temperature of the air that a person is in contact with, measured by the dry bulb temperature (DBT). This is often referred to as the main component of thermal comfort. The air temperature of a space is influenced by internal and external gains, as well as any means of ventilation and the building fabric.

(b) **air velocity**. The velocity of the air that a person is in contact with (measured in m/s). The faster the air is moving, the greater the exchange of heat between the person and the air (for example, draughts generally make us feel colder). Table 2.6 shows typical air velocities and their applicability regarding thermal comfort.

Section 2 – Fabric and comfort considerations

Table 2.6 Air velocity and thermal comfort

Velocity of air	Thermal comfort impact
0 m/s	Stationary air. (**Note**: minimum air change rates are required to maintain indoor air quality)
0.1 m/s	May be used as the assumed internal air velocity in some simple heat transfer calculations.
0.1–0.15 m/s and above	May be felt as a draught in a cold climate in the winter.
0.3 m/s and above	May be felt as a draught in a cold climate in the summer.
0.8–1 m/s and above	May be felt as a draught in a hot climate.

(c) **radiant temperature**. The temperature of a person's surroundings (including surfaces, heat generating equipment, the sun and the sky). This is generally expressed as mean radiant temperature (MRT, a weighted average of the temperature of the surfaces surrounding a person, which can be approximated by a globe thermometer) and any strong monodirectional radiation, such as radiation from the sun.

(d) **relative humidity (RH)**. The ratio between the actual amount of water vapour in the air and the maximum amount of water vapour that the air can hold at that air temperature, expressed as a percentage. The higher the RH, the more difficult it is to lose heat through the evaporation of sweat. A very low RH can cause comfort issues such as dry eyes or respiratory irritation. Recommended RH is typically 25–75 % for most indoor spaces and activities.

2.3.3 Personal comfort factors

The key factors affecting personal comfort are:

(a) **clothing**. This insulates a person from exchanging heat with the surrounding air and surfaces and affects the loss of heat through the evaporation of sweat. Clothing can be directly controlled by a person (they can take off or put on a jacket), whereas environmental factors may be beyond their control. Clothing is therefore deemed to be an adaptive comfort measure.

(b) **metabolic heat or level of activity**. Physical exercise generates heat and this heavily influences a person's thermal comfort. A stationary person will tend to feel cooler than a person who is exercising. Recommended design temperatures for exercise spaces will therefore be lower than those accommodating more sedentary activities.

(c) **wellbeing and sicknesses**, such as the common cold or flu, which affects our ability to maintain a body temperature of 37 °C at the core.

Other contributing factors include access to food and drink, acclimatization (this can be more difficult where there is a high outdoor-indoor temperature gradient) and an individual's general state of health. In addition, thermal comfort will be affected by whether a thermal environment is uniform or not. For example, draughts and heaters can create a scorched face/frozen back effect and hot feet/cold head and hands effect.

The thermal alliesthesia hypothesis goes beyond this and proposes that the hedonic qualities of the thermal environment (qualities of pleasantness or unpleasantness, or 'the pleasure principle') are determined as much by the general thermal state of the subject as by the environment itself. In its simplest form, cold stimuli will be perceived as pleasant by someone who is warm, whilst warm stimuli will be experienced as pleasant by someone who is cold. By introducing a spatial component to this, it can, for example, be pleasurable to wrap cool hands around a warm mug.

Section 2 – Fabric and comfort considerations

2.3.4 External design temperatures

To keep building occupants comfortable, heating systems are designed using the average near-extreme values of DBT in winter. This is not the lowest temperature recorded at a site, rather a weighted average based on historical data. Design temperatures do not only affect thermal comfort but also have a large influence on the capital cost of building services systems and some influence on running costs, and assessing the feasibility of heating technologies.

No single design temperature is given for a particular location, but there is a range from which a designer can select an appropriate temperature, bearing in mind the occupants' needs and the activity taking place on site.

Initial selection of an external design temperature should be based on meteorological data from the nearest recording stations to the site in question. However, a carefully selected design temperature should also consider microclimatic conditions such as altitude, proximity to the coast and urban heat island effects.

The Chartered Institution of Building Services Engineers' (CIBSE) Guide A: Environmental Design gives suggested design temperatures ranges for an array of locations in the UK; a typical winter design temperature could be around $-4\,°C$[10].

2.3.5 Internal design temperatures

The CIBSE's *Guide A: Environmental Design* also provides details of thermal comfort targets for different indoor spaces. Similarly to external design temperatures, suggested temperature ranges are given instead of specific values, since thermal comfort is the product of a range of factors, of which DBT is only one. The Guide also provides guidance on relative humidity, air movement velocity and radiant effects, influencing how design temperatures should be adjusted. For domestic applications target internal design temperatures are:

(a) bedroom 19 °C
(b) hall 19 °C;
(c) living room 22 °C;
(d) toilet 21 °C; and
(e) bathroom 22 °C.

It is noted that the 20 °C is used throughout in most assessment criteria, such as the SAP and the PHPP.

2.4 Heat loss calculations

2.4.1 Sizing a heating system

The correct sizing of a heating system is of paramount importance to its economical and reliable operation. If a system is undersized, it will not adequately meet its requirements in reaching and maintaining a comfortable set-point temperature, and its longevity will be affected due to prolonged strain on its components.

---

[10]Chartered Institution of Building Services Engineers (CIBSE) *Guide A: Environmental design* CIBSE, London, 2015. Available at: https://www.cibse.org/getattachment/Knowledge/CIBSE-Guide/CIBSE-Guide-A-Environmental-Design-NEW-2015/Guide-A-presentation.pdf.aspx

Section 2 – Fabric and comfort considerations

If a heating system is oversized, it will not operate optimally. Many oversized heat generators, such as boilers and heat pumps may suffer 'short cycling', which is where repeated starts throughout a day cause components to wear more quickly and in turn results in a reduced product lifespan.

It is important to maintain steady state conditions by installing a heating system with an output that matches or exceeds the amount of heat lost. Figure 2.7 shows an appropriately sized heating system that was determined using a heat loss calculation.

Figure 2.7 An appropriately sized heating system to maintain steady state conditions in a home

A heating system is sized using the total calculated losses (through fabric and ventilation) using a temperature differential defined by specified internal and external design temperatures, together with a size margin to account for conditions outside the design parameters (typically 10–15 %). As stated earlier, oversizing should be avoided and design conditions should account for a worst case scenario, where a building is required to heat to the set point from cold (i.e. equal to the external design temperature). In practice, this would be a rare and unlikely occurrence, since internal and external gains would provide for a higher ambient internal temperature.

Heat emitters are sized according to their rated output at a particular flow temperature in the context of a space's calculated heat loss. At conventional gas boiler flow temperatures (50–80 °C), heat emitters (typically radiators) can be smaller if flow temperatures are lower since the magnitude of heat delivered is greater at higher temperature and can be accomplished with a smaller emitter. However, solutions with heat pumps operate at lower temperatures (typically 35–55 °C) requiring larger surface area emitters that are capable of operating at lower temperatures.

Modern heating systems are often combined with domestic hot water (DHW) generation, which can contribute to the majority of the overall heating load. Where this is the case, there are considerations around anticipated simultaneous load and DHW storage that can influence the size of a heat generator.

Section 2 – Fabric and comfort considerations

2.4.2 U-value formula

$$\text{U-value} = \frac{1}{\left(\sum R + R_{\text{si}} + R_{\text{so}}\right)}$$

where:

$\sum R$ is the sum of layers R-values

R_{si} is the internal surface R-value $= 0.12$ m^{2}K/W

R_{so} is the external surface R-value $= 0.06$ m^{2}K/W

2.5 Procedure for heat loss calculation

2.5.1 Step 1: determine construction

Figure 2.8 shows a cross-section through a cavity wall.

Figure 2.8 Section through a cavity wall

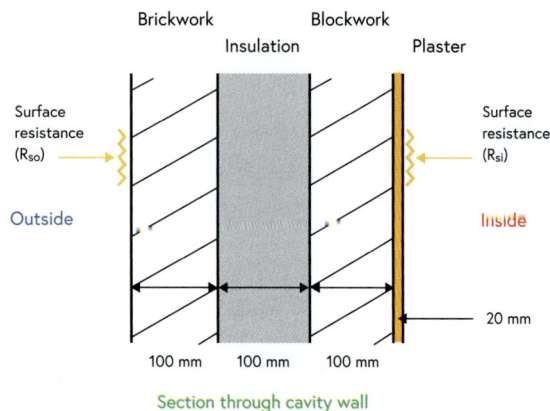

Section through cavity wall

2.5.2 Step 2: establish thermal conductivities

Thermal conductivities (obtained from manufacturer's information) include:

(a) brickwork $= 0.84$ W/mK
(b) insulation $= 0.03$ W/mK
(c) blockwork $= 0.65$ W/mK
(d) plaster $= 0.50$ W/mK
(e) $R_{\text{si}} = 0.12$ m^{2}K/W
(f) $R_{\text{so}} = 0.06$ m^{2}K/W

2.5.3   Step 3: calculate U-value

$$U = \cfrac{1}{R_{so} + \cfrac{l_{brick}}{\lambda_{brick}} + \cfrac{l_{insul}}{\lambda_{insul}} + \cfrac{l_{block}}{\lambda_{block}} + \cfrac{l_{plaster}}{\lambda_{plaster}} + R_{si}}$$

$$U = \cfrac{1}{0.06 + \cfrac{0.1}{0.84} + \cfrac{0.1}{0.03} + \cfrac{0.1}{0.65} + \cfrac{0.02}{0.5} + 0.12}$$

$$U = \cfrac{1}{3.83}$$

$$U = 0.261 \ \mathbf{W/m^2 K}$$

2.5.4   Step 4: determine heat loss

Total heat losses are used to determine the amount of heat required from a heat generator to reach and maintain the temperature setpoint within a building space. This is calculated by determining the fabric heat losses and adding them to the ventilation heat losses. The necessary formulae are:

For total heat loss:

$$Q_t = Q_{fabric} + Q_{vent}$$

For fabric heat loss:

$$Q_{fabric} = \Sigma(U \times A \times \Delta T)$$

For ventilation heat loss:

$$Q_{vent} = 0.33 \times n \times V \times \Delta T$$

where:

Q_t is the total heat loss

Q_{fabric} is the heat loss through fabric elements (W)

Σ is the sum

U is the U-value (W/m^{2}K)

A is the area (m^2)

ΔT is the temperature difference, or 'uplift': internal temperature to be reached (set point) minus external temperature (°C)

Q_{vent} is the ventilation heat loss due to infiltration and ventilation (W)

0.33 is the factor

n is the number of changes due to infiltration and ventilation

V is the room volume (m^3)

Section 2 – Fabric and comfort considerations

2.6 Example heat loss calculation

An example heat loss calculation for a small building is shown in Figure 2.9.

Outside design temperature = −4.5 °C
Inside set-point temperature = 20 °C
Building: length = 6 m; width = 3 m; height = 2.5 m
Window: length = 2 m; height = 1 m
Air change rate = 2 ACH

Figure 2.9 Example heat loss calculation for a small building

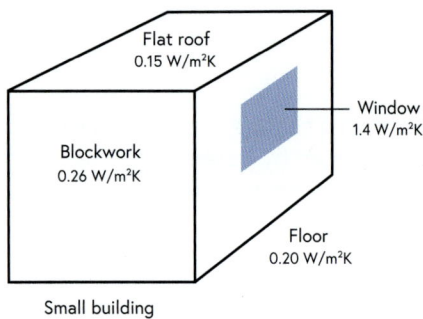

Small building

2.6.1 To calculate the heat loss of each element

$$Q_{\text{fabric}} = \Sigma(U \times A \times \Delta T)$$

For walls:

$$Q_{\text{win wall}} = 0.26 \times ((6 \times 2.5) - 2) \times (20 - -4.5)$$
$$Q_{\text{win wall}} = 82.81 \text{ W}$$
$$Q_{\text{opp wall}} = 0.26 \times 6 \times 2.5 \times (20 - -4.5)$$
$$Q_{\text{opp wall}} = 95.55 \text{ W}$$
$$Q_{\text{front and rear}} = 0.26 \times 3 \times 2.5 \times (20 - -4.5)$$
$$Q_{\text{front and rear}} = 95.55 \text{ W}$$
$$\mathbf{Q_{\text{total}} = 273.91 \text{ W}}$$

For window:

$$Q_{\text{window}} = 1.4 \times 2 \times 1 \times (20 - -4.5)$$
$$Q_{\text{window}} = 68.6 \text{ W}$$

For floor:

$$Q_{\text{floor}} = 0.2 \times 6 \times 3 \times (20 - -4.5)$$
$$Q_{\text{floor}} = 88.2 \text{ W}$$

For roof:

$$Q_{\text{roof}} = 0.15 \times 6 \times 3 \times (20 - -4.5)$$
$$Q_{\text{roof}} = 66.2 \text{ W}$$

2.6.2 To calculate ventilation heat loss

$$Q_{\text{vent}} = 0.33 \times n \times V \times \Delta T$$
$$Q_{\text{vent}} = 0.33 \times 2 \times 6 \times 3 \times 2.5 \times (20 - -4.5)$$
$$Q_{\text{vent}} = 727.65 \text{ W}$$

2.6.3 To calculate total heat loss

$$Q_{\text{t}} = Q_{\text{fabric}} + Q_{\text{vent}}$$

Blockwork: 273.9 W

Window heat loss: 68.6 W

Floor heat loss: 88.2 W

Roof: 66.2 W

Q_{fabric}: 496.9 W (or 0.50 kW)

Ventilation: 727.65 W

$Q_{\text{t}} = 1150.3$ W (or 1.15 kW)

Total heat loss is often normalized by floor area:

$$\frac{1150.3 \text{ W}}{18 \text{ m}} = 63.9 \text{ W/m}^2$$

This example shows that ventilation is a major source of heat loss.

2.7 The performance gap

It has been recognized that there is often a significant 'gap', where the actual thermal performance of a building is different to the predicted performance. This is particularly crucial in terms of heating system design, where it is critical to size the heat generators and emitters to ensure the correct supply of heat for ideal operating conditions to achieve optimum efficiency.

To size the heating system effectively, it is important to measure the thermal performance – the heat transfer coefficients (HTC) or HLP – of the dwelling using a method that accurately represents the thermal performance of the building, whether it is a new-build or an existing building. The thermal performance of buildings has been shown to vary significantly from design expectation (with variations of >100 % in some cases) and these variations can significantly undermine heating system designs.

The HLP of apparently similar buildings can be very different, for example in measurements of 10 houses of the same archetype on the same site, observed differences from the design thermal performance can range from −37 % to +44 % (see Figure 2.10). This shows that if a heating system for each house was sized based on design assumptions alone, there could be major shortcomings in the heating system performance.

2.8 Thermal modelling software

In practice, software is used to carry out and automate the calculation of many engineering tasks. The software capabilities for building physics have developed significantly. They can now more accurately assess heat losses and gains, thermal comfort considerations (including PMV and PPD) and overheating risk using dynamic simulation modelling with bespoke weather files. These software packages contain detailed hourly information on many data points, so that their capabilities extend to all aspects of building physics. Examples of software packages include:

(a) Virtual Environment (Integrated Environmental Solutions): full thermal modelling suite offering dynamic simulation for HVAC loads calculation, thermal comfort, energy compliance, costing and carbon calculations;
(b) Tas (Environmental Design Solutions Limited): full thermal modelling suite offering dynamic simulation for HVAC, thermal comfort, compliance and carbon calculations;
(c) Bentley (Hevacomp): building energy analysis software;
(d) Psi-Therm: bespoke complex thermal bridging calculation; and
(e) Design SAP (Elmhurst Energy) domestic building regulations software incorporating accurate U-value calculation tools.

Figure 2.10 Observed variance in HLPs between similar buildings

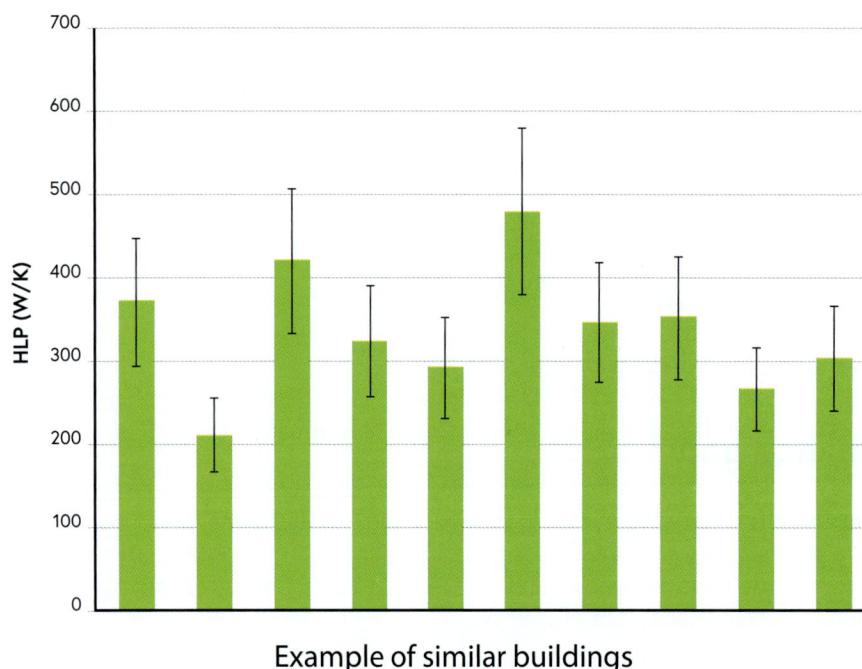

Example of similar buildings

2.9 Summary

(a) Wherever possible plan and design for free heat from the sun using south-facing glazing to provide as much passive winter heating as possible. This will reduce electric heating loads.
(b) Mitigate unwanted summertime solar gains through the provision of external shading from fixed and moveable blinds/shades or planting suitable trees and shrubs. This can negate the need for summertime cooling.
(c) Insulate and install high performance glazing. Insulation significantly reduces heating demands and minimizes electric heating capacity and running costs.
(d) Keep the building airtight by sealing up all leakage paths, but plan for controllable ventilation with heat recovery.
(e) Aim high in new builds and, where possible, adopt the Passivhaus standard.
(f) Upon completion, monitor the consumption of the building and compare against the design stage estimates and assess any performance gaps.

▬ Section 3

Heat pumps

The most efficient form of electrified heating is the heat pump because, in the right conditions, its heat output can be several times its energy input. Whilst heat pumps have been commonly used to provide heating and cooling in commercial buildings over the past 30 years, it is only in the past 5–10 years that they have been installed to provide heating in UK domestic properties. The uptake of the technology in EU member states has been greater, with an estimated total of 40 million heat pumps installed to date.

This Section specifically considers heat pumps as a form of domestic electrified heating and explores how heat pumps work, what types of heat pump are available, the key operating factors associated with heat pumps, how heat pumps and efficiency ratings are classified and some of the practicalities to consider when installing and operating a heat pump.

3.1   Heat pump theory

Simply put, a heat pump is a device that transfers energy in the form of useful heat from one medium to another (see Figure 3.1). This is not a new concept and has been used for many decades since the invention of refrigeration. This technology is more commonly associated with chilling processes rather than heating, but the refrigeration cycle simply moves heat from one medium to another by extracting a large quantity of low grade, low temperature heat, concentrating it and depositing it at a higher temperature for use elsewhere.

In buildings this cycle can extract heat from a range of media including air, ground or water and transfer it to another media, such as air or water, depending on the emitter and delivery mechanism.

Figure 3.1   The principal components of a heat pump

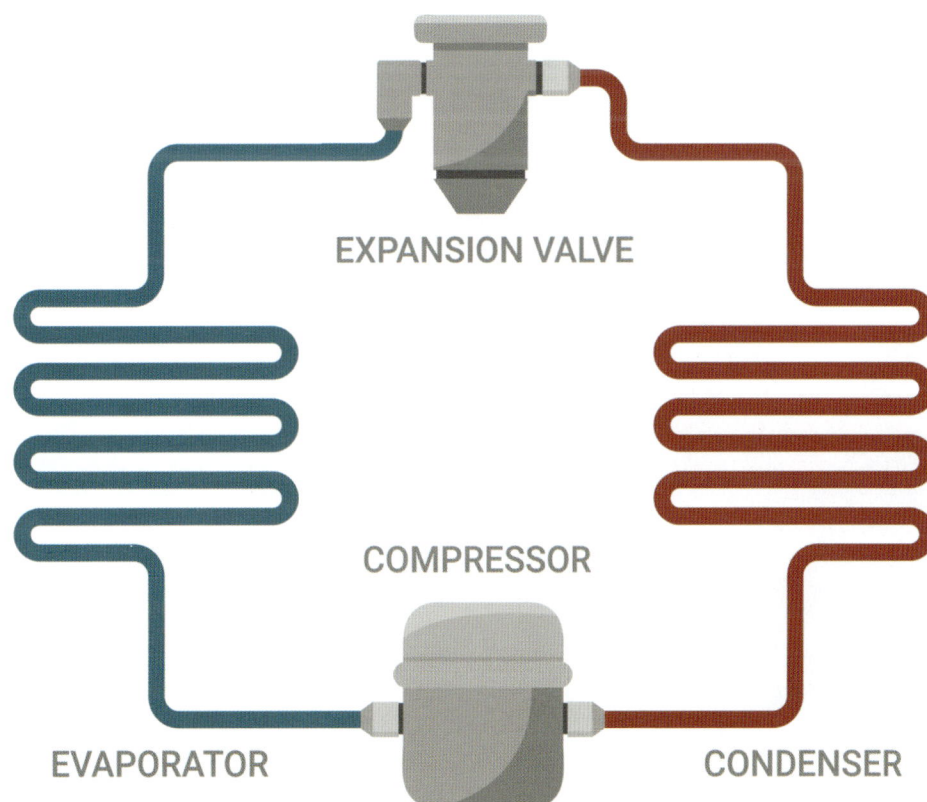

EXPANSION VALVE

COMPRESSOR

EVAPORATOR

CONDENSER

Section 3 – Heat pumps

The European standard for testing and rating heat pump performance DIN EN 14511-1 Part 1 *Air conditioners, liquid chilling packages and heat pumps for space heating and cooling and process chillers, with electrically driven compressors* defines a heat pump as 'an encased assembly or assemblies designed as a unit to provide the delivery of heat that includes an electrically operated refrigeration system for heating'. A heat pump can cool, circulate, clean and dehumidify the air (cooling is achieved by reversing the refrigeration cycle).

There are various types of heat pump that transfer heat between a combination of media, such as air to air, air to water, water to water, water to air or ground to water.

3.1.1 Types of heat pump

Heat pumps are classified according to the type of media through which they transfer heat. The source medium can be either air or water and the heat can be transferred to either an air or water-borne delivery system inside a building.

There are three main types of heat pumps (both ground and water source heat pumps extract heat from fluids).

Air source heat pumps

In the UK, the most common heat pump for use in most domestic and non-domestic buildings is an air source heat pump (ASHP). This comprises of an outdoor unit which uses a large fan to forcibly pass air through an evaporator, where heat is absorbed into a refrigerant. This is then compressed, raising its temperature and it is then transmitted inside the building. In most cases, this heat is transferred through a heat exchanger into a separate circuit to be distributed around the building, either via the air through ducts or more commonly through water to radiators or underfloor pipes (see Figure 3.2). An ASHP can typically deliver efficiencies of up to 400 % (a coefficient of performance (CoP) of 4.0).

Figure 3.2   An ASHP providing heat via water-filled radiators (also known as an air to water heat pump)

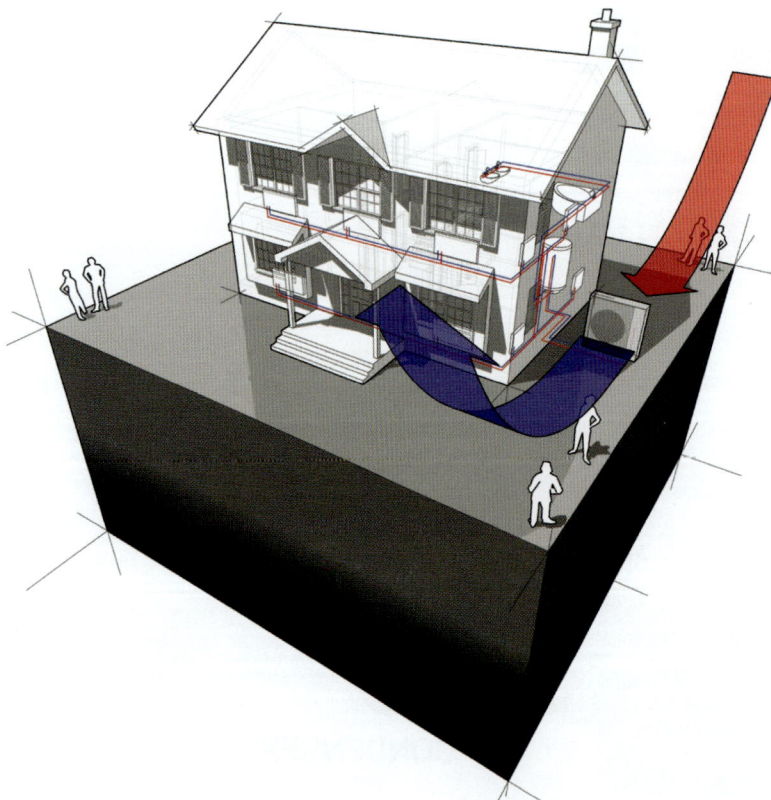

Section 3 – Heat pumps

The key advantage of an air source system is its relative simplicity and lower cost (compared with ground and water source heat pumps). However, its major drawback is the highly fluctuating nature of the source medium (ambient external air temperatures), which means the efficiency of the process is dependent on the ambient temperature.

Ground source heat pumps

Ground source heat pumps (GSHPs) are available in two types: they can either extract heat via a long loop of pipes that absorb heat from the surrounding earth (known as 'closed loop') or via ground water pumped from deep underground (known as 'open loop'). Because GSHPs use fluids as their source medium, they are essentially a type of water source heat pump. A GSHP can typically deliver efficiencies of up to 600 % (a CoP of 6.0).

(a) Closed loop. The source heat exchanger can be either shallow or deep. In shallow applications, coiled or horizontal parallel pipes are laid at a depth of around 2–3 m underground across a large expanse (typically twice that of the required heated floor area). In deep applications, boreholes are drilled around 50–150 m into the ground and a U-shaped loop is lowered into the hole. This is then backfilled with sand or another fine medium. The pump is required to slowly circulate a mixture of water and antifreeze around the loop (see Figure 3.3).

(b) Open loop. Water is abstracted from a deep borehole using a pump and passed through a heat exchanger with the evaporator to absorb heat energy. Once heat has been extracted, the now cooled water is discharged back into another borehole, or to a water course (ensuring suitable distance from the abstraction borehole to avoid mixing).

In both cases there is no need for an external unit as no external air supply is required. Heat pumps are typically situated in a utility cupboard or plant space. In a closed loop system, the cooled transfer medium is then pumped around the loop again to absorb more heat, whereas in an open loop system, the cooled water is discharged.

Figure 3.3   A closed loop GSHP using a shallow excavated ground area and coiled pipes

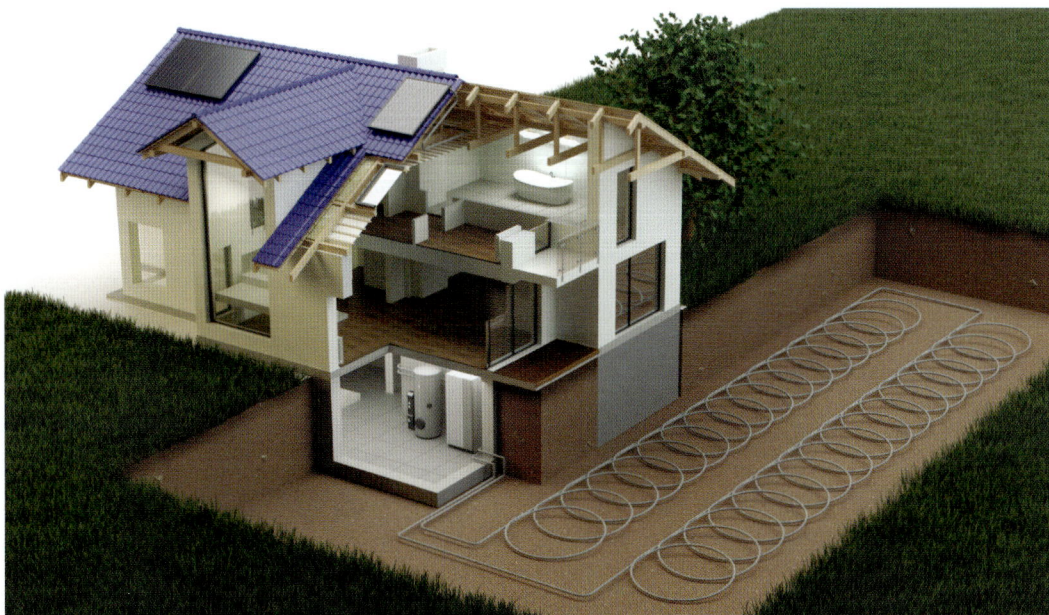

Although GSHPs are less common than ASHPs, they offer more stable source temperatures than air, which means they are more reliable and have a constant efficiency compared with an ASHP – particularly in winter months when demand for heat is highest. In the UK, shallow ground temperatures of 2–3 m typically fluctuate by around 10 °C throughout the year, and at 100 m (for geothermal boreholes) by no more than 5 °C. Comparatively, ambient air temperature can fluctuate by around 40 °C throughout the year, with the lowest temperature inconveniently coinciding with the highest heat demand.

The major drawback of GSHPs is their installation cost: surveys, excavations and boreholes can amount to more than double the price of installing an ASHP. They are also subject to many suitability criteria, ranging from site access to underlying geology and proximity to water courses.

Water source heat pumps

Water source heat pumps (WSHPs) are less common than ASHPs and GSHPs but adopt broadly the same principles as GSHPs. Water is either abstracted directly from a water course and discharged down-stream, or coiled loops are immersed into a large body of water. These carry many of the advantages of GSHPs, but water temperature fluctuations are more variable depending on the depth of installation and the size and refresh rate of the water course or body. Also, smaller bodies of water could be susceptible to localized freezing that could render the heat pump inoperable. A WSHP can typically deliver efficiencies up to 600 % (a CoP of 6.0).

For a WSHP to be viable, the property must be within a reasonable distance of the water course or body to avoid transmission heat losses. Permission must also be sought from the appropriate agency ahead of installation. For open-loop systems, a licence must be obtained to allow water to be abstracted from and/or discharged into a water course. There are also several specific design and maintenance considerations for open-loop systems, such as corrosion protection (particularly where sea water is used) and systems employed to prevent filtration blockage.

3.1.2 Principal heat pump components

A heat pump system typically comprises the following components:

(a) **evaporator**. This is the heat exchanger where refrigerant fluid is evaporated and heat is absorbed from the surrounding medium (air, water, or ground), reducing its temperature. These are typically located outside the building, particularly if the heat pump is an ASHP (see Figure 3.4).
(b) **expansion valves**. These are flow-restricting devices that cause a sudden drop in pressure of the working fluid. They also separate the high-pressure and low-pressure sides of the heat pump system.
(c) **compressor**. The compressor is the mechanical driver of the heat pump system where the input energy (electricity) is expended. Compression raises the temperature of the refrigerant vapour so that it can subsequently be condensed back into a liquid state to resume the heat transfer process. It also pumps the refrigerant fluid through the circuit. It is the compressor that generates the noise of the system and it is usually located in the outdoor unit.
(d) **interconnecting pipes transporting refrigerant**. The refrigerant pipework in the circuit must be well sealed to avoid leaks (and the release of any harmful substances) and to maintain an efficient operation.
(e) **condenser**. The heat exchanger is where hot, compressed refrigerant gas is condensed to a liquid and further cooled; heat is rejected (released) where it needs to be used, usually inside a building in a floor-standing water-filled tank via a heat exchanger. This is typically located in a cupboard along with the associated pipework, pumps and controls.
(f) **controls/programmer**. Effective controls govern the heat pump's operation to ensure it delivers heat efficiently, initiates defrost cycles and identifies faults. The controls also operate the heating system's emitters and governs the room temperature.

Section 3 – Heat pumps

Figure 3.4 An ASHP evaporator located outside a home

These components are arranged in the vapour compression cycle as shown in Figure 3.5.

Figure 3.5 The four stages of the vapour compression cycle (showing principal components of a heat pump)

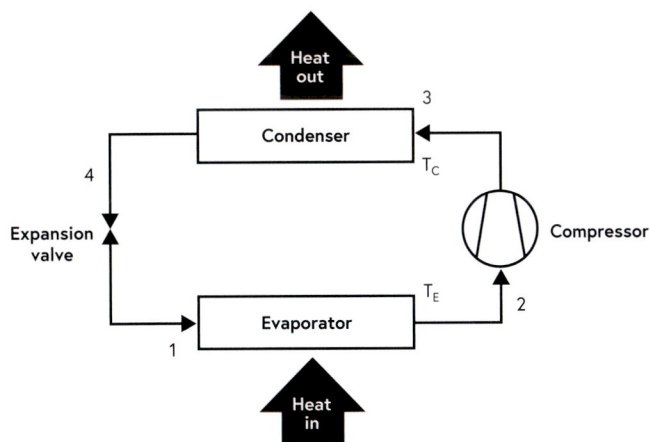

where:

T_C is compressor temperature

T_E is evaporator temperature

Note: In a cooling cycle, this is reversed and the roles of the heat exchangers (condenser and evaporator) are inverted.

Section 3 – Heat pumps

3.1.3 Heat pump cycle

Energy moves in a heat pump by absorbing energy from outside through an evaporator and rejecting energy through a condenser. This is known as the vapour compression cycle, whereby the latent heat of the refrigerant is used to either absorb or release energy as it makes a transition from a liquid to a gas state or vice versa. Figure 3.6 shows the vapour compression cycle as a temperature-heat diagram.

Figure 3.6   Temperature-heat vapour compression cycle

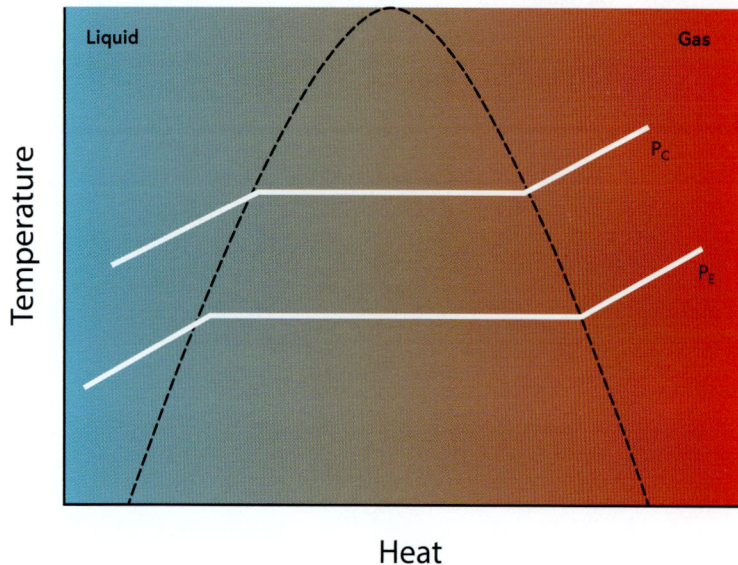

where:

P_C is compressor pressure

P_E is evaporator pressure

In the liquid state, an increase in heat results in a proportionate increase in temperature as the liquid absorbs the heat. The refrigerant then starts the transition from liquid to gas (shown by the dashed black line) and there is no change in temperature as heat is increased. This is because all the energy is required for phase transition. When this transition is complete the refrigerant is in a gas (or vapour) state and an increase in heat returns a proportionate increase in temperature.

If heat is released, the temperature of the refrigerant falls until it reaches the saturation line. It then enters phase transition and changes from gas to liquid. During phase transition, there is no change in temperature until it becomes a liquid and the temperature again decreases in proportion to the heat energy being released.

The temperature at which phase transition occurs is determined by the refrigerant pressure. If the pressure is increased, for example from P_E (evaporator pressure) to P_C (compressor pressure), the transition occurs at a higher temperature.

A vapour compression cycle can also be represented by a pressure-heat diagram (sometimes referred to as a Ph diagram). Figure 3.7 shows a simplified pressure-heat diagram with lines of constant temperature, T_E (evaporator temperature) and T_C (compressor temperature). When the refrigerant is either in a liquid or a gas state, heat is neither absorbed nor released with changes in pressure. However, when the refrigerant is in phase transition (shown by the area within the black dashes), it absorbs or releases heat at a constant temperature and pressure.

Section 3 – Heat pumps

Figure 3.7   Pressure-heat vapour compression cycle

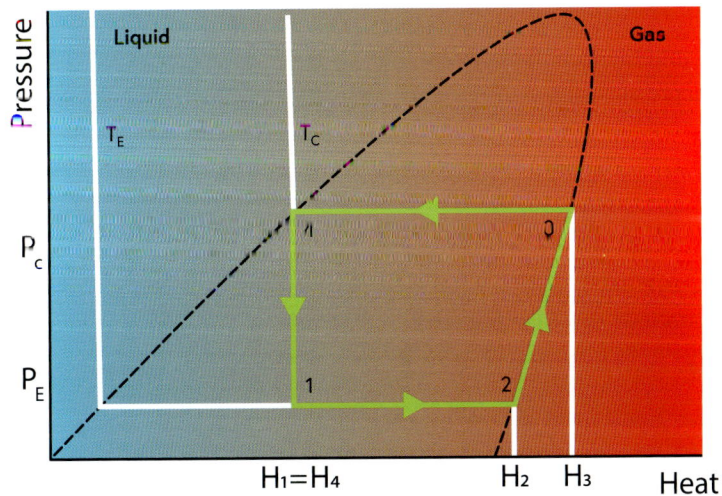

where:

T_C is compressor temperature

T_E is evaporator temperature

P_C is compressor pressure

P_E is evaporator pressure

Superimposed on the pressure-heat diagram is the vapour-pressure cycle, which is represented by the green arrows. This ignores pressure and temperature losses and the thermodynamic properties of the refrigerant that will influence the cycle. The stages of the cycle are detailed in Table 3.1.

Table 3.1   Vapour-pressure cycle

Cycle point	Description
1	The refrigerant is a liquid and enters the evaporator at a pressure P_E and a temperature T_E and this corresponds to the heat source, e.g. water or air.
1–2	The refrigerant passes through the evaporator and its temperature and pressure initiates phase transition with the refrigerant absorbing heat from the heat source.
2	The refrigerant is a gas at pressure P_E and temperature T_E.
2–3	The refrigerant is compressed increasing its pressure to P_C and its temperature to T_C.
3	The refrigerant is a gas at pressure P_C and temperature T_C.
3–4	The refrigerant passes through the condenser and its temperature and pressure initiates phase transition with the refrigerant releasing heat to the area to be heated.
4	The refrigerant is a liquid at pressure P_E and temperature T_E.
4–1	The refrigerant passes through the expansion valve, which reduces its pressure to P_E and temperature to T_E, ready for the cycle to be repeated.

Section 3 – Heat pumps

The efficiency of the cycle, the heat pump's CoP (see Section 3.3.3), is expressed as:

$$\frac{\text{Heat released into the area to be heated}}{\text{Compressor energy}} = \frac{H_3 - H_4}{H_3 - H_2}$$

3.1.4 Refrigerants

A refrigerant is a man-made substance or mixture, usually a fluid, used in a heat pump that is sensitive to fluctuations in pressure, and these substances provide the work in a refrigeration cycle. Refrigerants transition from liquid to gas, and back again, to absorb and release heat energy.

Many types of these working fluids have been used over the years, but many are now being either banned or phased out due to associated harmful properties. For example, some produce gases that possess powerful global warming potential (GWP) properties that, when released into the atmosphere are up to 23,000 times more harmful than CO_2. This is not simply due to these gases' ability to cause a greenhouse effect, but is a combination of this and their ability to deplete the ozone layer protecting the earth from harmful ultra-violet radiation emitted by the sun. Therefore, GWP is a holistic indication of a gas's ability to effect increased global warming.

In Europe, fluorinated gases (F-gases) are controlled using F-gas regulations. F-gases include hydrofluorocarbons, perfluorocarbons and sulphur hexafluoride. F-gas requirements include restrictions on the availability of certain refrigerants with a high GWP, refrigerant leakage checks, record keeping and accreditation schemes for those involved in heat pump maintenance.

Refrigerants that have been phased out include R-22 (GWP 1,810) and R-407c (GWP 1,732). Other refrigerants such as R-410a (GWP 2,088) are being phased out with a target date of 2030. Very low GWP solutions include R32 (GWP 675), ammonia, CO_2 and hydrocarbons, R-1234yf, R-1234ze and R1233zd.

Different refrigerants possess different preferable properties dependent on the desired operating factors, such as their boiling/liquification point, useful lifespan or flammability. In heat pumps, the type of refrigerant heavily influences its ability to transfer heat efficiently.

3.2 Heat pump operating factors

3.2.1 Operating conditions for heat pumps

A key limitation of heat pumps' viability compared with conventional heating technologies is their restricted operating conditions. The operation of fossil-fuelled boilers and electrical resistance heating are largely unaffected by external conditions, whereas heat pumps have set limiting parameters determined by the ambient temperature of the source medium.

Some heat pumps are theoretically able to extract heat energy from sources at temperatures as low as $0\,°K$ (or absolute zero, $-273\,°C$). However, the amount of input energy required to extract and distribute heat below a certain ambient temperature becomes uneconomical given that resistance heating is 100 % efficient. If more electrical energy is needed for the process than heat energy generated, then the heat pump is not viable.

Section 3 – Heat pumps

A heat pump's efficiency is also determined by the temperature differential between the source and rejection media, and for most heat pumps, the economical output temperature is between 30 °C and 55 °C, depending on the temperature of the source medium. The lower the delta-T (the temperature differential between the source and output temperature), the more efficient the heat pump cycle will be. However, the UK's moderate temperate climate generally presents favourable conditions for heat pump operation.

3.2.2 External ambient operating temperature

The optimal climates for heat pumps are warm or moderate, as the availability of heat in the ambient air is directly related to operation efficiency.

The ambient temperature at which heat pumps can operate efficiently varies between manufacturers, makes and models, but an optimal high-performance temperature for heat pumps is around 7–10 °C. For most conventional heat pumps, the lower limit of operation is around –15 °C. Heat pumps can operate at cold temperatures (see Figure 3.8), but only when correctly selected.

Figure 3.8   A heat pump installed in a cold climate

Section 3 – Heat pumps

In the UK, the mean annual air temperature is around 9–11 °C and is consistent with the year-round ground temperature at 6 m depth. This means that heat pumps are generally well suited to the UK climate, but GSHPs are particularly favourable due to the relative stability of the ground temperature.

3.2.3 Heat pump internal output temperatures

The output temperatures for heat pumps can vary depending on the external ambient conditions, heat pump type and refrigerant used. For the most part, heat pump output temperature is most efficient at a low-grade heat of around 35–45 °C, which works well in underfloor heating and correctly sized radiators (see Figure 3.9). In certain situations, fan-assisted radiators may be appropriate using low grade heat, particularly if space is restricted for larger radiators.

Figure 3.9 A heat pump used for a radiator and underfloor heating

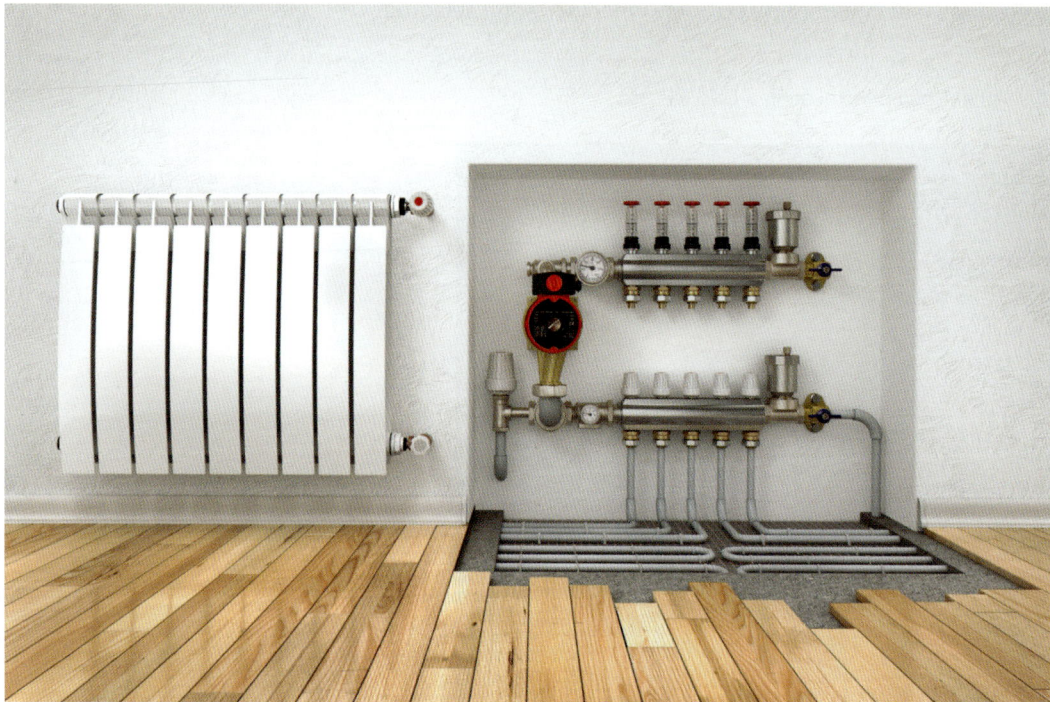

However, flow temperatures of up to 55 °C can also be provided efficiently. It should be noted that in existing installations, radiators will generally need to be oversized compared with boiler-generated heating, given that this is normally provided at 70–80 °C. Should this not be practically possible, high-temperature heat pumps can deliver hot water at up to 80 °C but are normally used for larger applications and do not operate efficiently at lower loads. High temperature heat pumps can therefore be considered in district or site scale installations for the provision of heat to a large number of dwellings, or as part of a mixed-use development.

3.2.4 Defrost cycles

Operating at low temperatures (i.e. between –2 °C and 2 °C) requires defrost cycles (typically 1–2 per hour). This is when the system reverses for a short period to provide heat to clear the evaporator coils.

Section 3 – Heat pumps

As the temperature falls further no defrosts are normally necessary as most of the moisture has been removed from the air as frost.

ASHPs are continually extracting energy from the outside air, causing moisture to condense on the evaporator. In normal circumstances, this falls off the evaporator and is drained away. As ambient air temperature falls however, any condensation present on the evaporator will cause frost to form (see Figure 3.10). If this is allowed to accumulate, it will hinder the heat pump's ability to absorb heat into the refrigerant, degrade its efficiency and cause the cycle to fail

Figure 3.10   A poorly operating heat pump with an accumulation of frost

To counter the accumulation of frost, the heat pump will automatically enter a defrost cycle, where the heat pump operated in reverse. This continues until no frost or ice remains, so that the evaporator can absorb heat effectively.

The stages of the process are:

(a) a refrigerant diverting valve triggers the reverse-cycle operation;
(b) compressed vapour from the compressor enters what was the evaporator in heating mode (now acting as the condenser);
(c) as the refrigerant vapour condenses to liquid, the rejected energy will be used to defrost the heat exchanger;

(d) heat energy is absorbed from the primary water circuit in what was the condenser in heating mode (now acting as the evaporator) to boil off the liquid refrigerant;

(e) the defrost will last between 1–2 minutes and can occur around once or twice per hour;

(f) low temperature water vapour will be seen coming off the unit during the defrost operation; and

(g) the primary water circuit will experience a small temperature drop and heat emitters may become cooler to the touch.

During this time, the fan will be turned off and a direct electric auxiliary heater can be used to provide continuous heat inside the building.

3.2.5 Sizing considerations

Similarly to any heating system, a heat pump should be sized based on the total heat losses of the building (both fabric and ventilation), and to satisfy DHW demand if applicable. The correct sizing of heat pumps is an important factor as both undersizing and oversizing can result in problems in terms of efficiency, reliability and economic operation.

Undersizing

Not matching the heat load with the heat pump output can cause problems. As ambient temperature falls (and the delta-T rises), so does efficiency and in turn heat output (kW capacity). As less heat is extracted from the air, the compressor works harder to overcome peak heat load, resulting in reduced efficiency and energy consumption.

Oversizing

Selecting a larger output heat pump than required can also cause operational issues.

Whilst heat pumps are provided with inverters to modulate the compressor speed (the minimum capacity, or 'turndown', is typically around 30 %). When the heat output required is less than 30 %, the compressor switches on and off and 'short-cycling' occurs. This not only wastes energy but causes a reduction in compressor life, since most wear occurs when the compressor is started. For this reason, the anticipated life cycle of a heat pump is sometimes quoted in number of starts, as opposed to years of operation.

If a larger heat pump is required, a solution to oversizing can be to add an appropriately sized thermal store or 'buffer vessel' to the heating and/or DHW system. This provides more continuous work for the heat pump and thermally charges a volume of water, which can be fed over time into the heating system. This also resolves issues with simultaneous demand, where a heat pump may struggle to cater for a sudden increase in demand, for example if several appliances are switched on at once whilst the space heating is also operating.

3.3 Heat pump efficiency

3.3.1 Heat pump efficiency measures

There are many factors that influence the efficiency of heat pumps, governed mainly by the temperature differential between the source and output media (delta-T), but there are other factors that influence efficiency.

Section 3 – Heat pumps

Methods of increasing efficiency in heat pumps include:

(a) combining heating and cooling operations using thermal storage, where rejected heat can be deposited into a heat sink during cooling season and recovered during the heating season. This can typically be achieved using the ground in larger, deeper installations but is dependent on ground conditions. This raises the efficiency of both cooling and heating operations.

(b) recovering waste heat. Heat rejected from external processes, typically those with cooling operations, can be used to maximize heating efficiency.

In addition, the design of heat pumps can include invertors that achieve variable flow of the refrigerant using variable speed drives to control energy transfers and energy inputs. The amount of energy used by a pump or fan is proportional to the square of its speed, for example, if a compressor is operated at half its rated speed, it will consume a quarter of the energy required at full load. This is governed by performance curves and can only operate to a certain turndown ratio.

3.3.2 Heat source/sink influence

To increase heat pump efficiency, the delta-T must be brought to the lowest possible differential value. This means that the source must be made warmer or the sink (the internal building set point or DHW temperature) must be made cooler. Since it is not reasonable to lower the output temperature beyond what is useful (i.e. 35 °C for space heating or 50 °C for DHW), the source must be selected to provide an operation that is as efficient as possible.

Using the ground or a body of water as the source and/or sink maintains a more stable temperature than air, thus presenting a heat pump with a more constant delta-T throughout the season, regardless of the external ambient temperature. This is the main advantage that ground and water source have over air source and can yield efficiency benefits of up to 50 %.

3.3.3 Coefficient of performance

The CoP of a heat pump is a ratio of useful heating provided to work input at a single point in time. This is given as a single figure and can be quoted as kW/kW, or specifically kW thermal/kW electrical (kW_{th}/kW_e). This is equivalent to the efficiency quoted for combustion heaters, which is usually quoted as a percentage.

The higher the CoP, the higher the efficiency and the lower operating costs will be. To be competitive with gas on fuel cost, a heat pump must operate with an average CoP of more than 3.0, but in the future this cost will need to be compared with other options when the government bans the use of natural gas in line with decarbonization measures.

$$\text{CoP} = \frac{\text{Rate of heat output at condenser (kW)}}{\text{Power input (kW)}}$$

$$\text{CoP} = \frac{Q_{out}}{W_{cycle}}$$

The heat delivered (Q_{out}) is in kW.

The power input to the unit (W_{cycle}) refers to the total electrical input power of all components (most typically the compressor) of the unit measured in kW.

An example CoP calculation is shown in Figure 3.11.

Figure 3.11   Example CoP in a heat pump with a 1 kW rated compressor that achieves 4 kW of heating

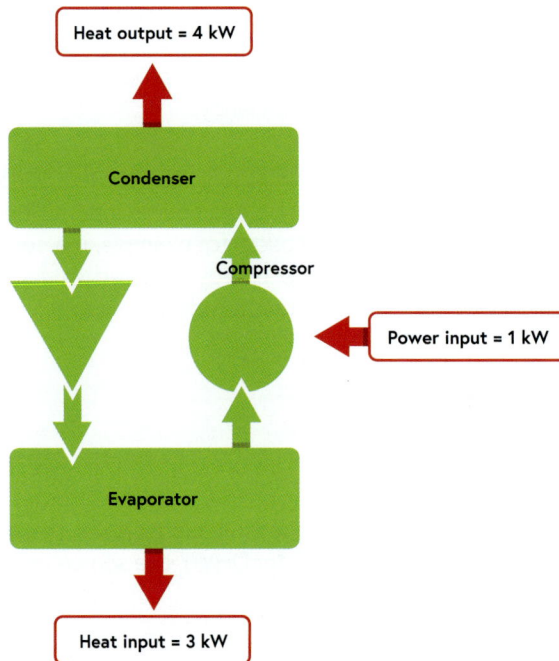

$$CoP = \frac{\text{Rate of heat output at condenser}}{\text{Power input}}$$

$$CoP = \frac{4 \text{ kW}}{1 \text{ kW}} = 4$$

3.3.4   Typical CoPs

Given that the efficiency of a heat pump largely is dependent on the differential between its input and output temperatures (delta-T), the temperature stability of the source medium is highly influential in determining its seasonal CoP.

ASHP (typical CoP 3.0–4.0)

Whilst outdoor ambient temperatures can reach more than 30 °C in the UK, this occurs when demand is lowest (i.e. heating is not required). Most heating is required when the outdoor ambient temperature is below 15 °C and fluctuates highly on a daily basis. Therefore, ASHPs offer the lowest CoPs.

GSHP (typical CoP 4.0–6.0)

GSHPs offer more consistent CoPs, due to the stable nature of their source media. Whilst CoPs are comparatively lower during summer months for GSHPs than ASHPs, they are higher during winter when

Section 3 – Heat pumps

there is a higher demand for heat. The temperature of the ground becomes more stable with depth, so a deep borehole will yield better CoPs than shallow ground collectors.

It should also be noted that if sized incorrectly or collectors are too concentrated within a small area, there is a risk of localized freezing if a system operates as a heating-only system

WSHP (typical CoP 4.0–6.0)

Like GSHPs, WSHP CoPs are dependent on the annual fluctuation of the source media. If a collector is situated within a large body of water, its temperature will be more stable resulting in a higher CoP. If open source, the CoP will be determined by the seasonal temperature fluctuation. This is likely to be lower than ASHPs, however, since water is less prone to fluctuation.

WSHPs are prone to the same limitations as GSHPs, as a large collector area within a small body of water could lead to localized freezing, which will have a detrimental effect on CoP and the system's ability to operate.

3.3.5   Seasonal CoP and seasonal performance factor

The coefficient of performance is a snapshot in time and indicates peak performance. The seasonal CoP (SCoP) is a figure used to describe a heat pump's theoretical expected performance in a given environment and is a weighted average based on expected proportion of time running at part and full load for a specific delta-T. This will be given as a single figure for a specific flow temperature in each region. Therefore, a SCoP given for a flow temperature of 35 °C will typically be better than the same heat pump outputting a flow temperature of 55 °C. Similarly, a heat pump being marketed for sale in the UK will typically have a higher quoted SCoP than the same one marketed in, for example, Norway.

Given that SCoPs give a measure of average efficiency, it is this figure that can be used to compare different system models and against conventional heating devices. SCoPs are also required for Building Regulations Part L compliance calculations.

For large projects, it is usual for a manufacturer to perform a bespoke calculation to determine the SCoP in an exact location using expected usage data.

The seasonal performance factor (SPF) is a better and more accurate measure of how efficiently a heat pump is operating. The SPF is a measure of the operating performance of an electric heat pump heating system over a year. It is the ratio of the heat delivered to the total electrical energy supplied over the year.

The SPF is calculated using the formula:

$$SPF = \frac{\text{Total heat energy output per annum (kWh)}}{\text{Total input electricity per annum (kWh)}}$$

3.3.6   The Energy-related Products Directive

The Energy-related Products Directive (ErP 2009/125/EC) is a key part of the European Union's drive to encourage consumers to use more energy-efficient products. However, a similar scheme is likely to be developed by the UK government.

The ErP 2009/125/EC is a comprehensive legislation that covers 'any product that uses, generates, transfers or measures energy, whether electricity, gas or fossil fuel'. Heat pumps fall into the scope of the ErP 2009/125/EC, as energy consuming and transferring equipment.

Section 3 – Heat pumps

The ErP 2009/125/EC is a two-part strategy:

1. **ecodesign regulations**: this requires manufacturers to produce products that meet stringent minimum performance standards; and
2. **energy labelling regulations**: products must be clearly labelled (see Figure 3.12) using a standard methodology so that consumers can quickly understand the energy efficiency of the products they purchase.

Figure 3.12   An example ErP 2009/125/EC energy label

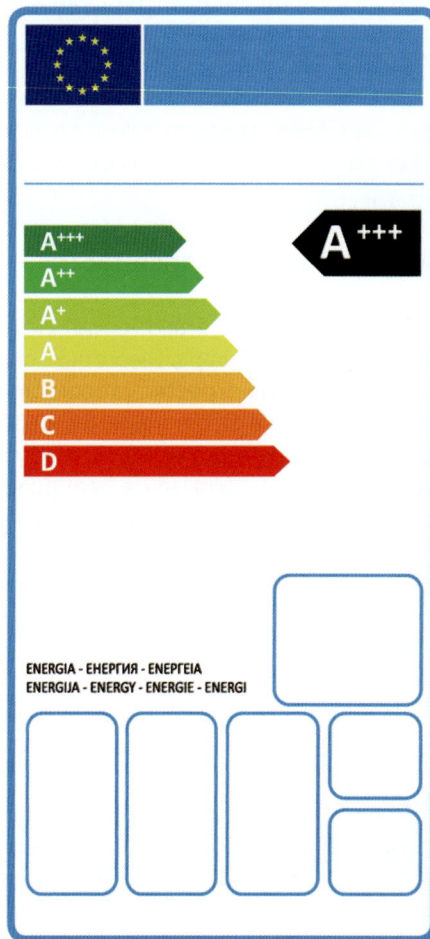

3.3.7   Ecodesign

The move to seasonal space heating energy efficiency as the main indicator of the energy efficiency of a product is significant. Previously, manufacturers often stated their products' efficiency at a single point in time. This often provided a theoretical maximum efficiency that was distorted across the year. The advantage of the ecodesign regulations for heating equipment is the requirement to report product performance in terms of a more realistic seasonal energy efficiency measure.

Seasonal space heating energy efficiency (SSHEE) is defined in the ErP 2009/125/EC legislation as, 'the ratio between the space heating demand for a designated heating season, supplied by a heater and the annual energy consumption required to meet this demand, expressed in %'.

Section 3 – Heat pumps

Table 3.2 sets out the ecodesign SSHEE minimum standards for each heating technology:

Table 3.2   Ecodesign SSHEE minimum standards

Space heating technology	Minimum SSHEE
Electric boiler space heaters and electric boiler combination heaters	35 %
Cogeneration space heaters	100 %
Heat pump space heaters and heat pump combination heaters	110 %
Low temperature heat pumps	125 %

3.3.8 Energy labelling

Energy labels help consumers to make energy efficient choices. Labels provide consumers with clear information on product performance to enable comparisons. For example, when considering space heaters, the energy efficiency labels run from G (lowest) to A+++ (highest). Typically, a gas combi boiler can achieve an A rating, traditional heat pumps a+ rating, invertor heat pumps an A++ rating and future heat pump technology (not yet developed) an A+++.

Heat pump labels also show a European temperature map displaying three indicative temperature zones. This is important for heat pumps as performance can be affected by climate. As well as a label, each product has a 'fiche', which contains more detailed information on how the classifications of the label have been achieved.

3.4 Heat pump practicalities

3.4.1 Practicalities of heat pumps

Besides technical considerations around heat pumps, there are several practical considerations around their installation and continued operation which can affect their suitability for a particular situation. These include:

(a) **air flows**: ASHPs require minimum clearances to the rear, sides and front;
(b) **access**: for maintenance and isolation in case of fault;
(c) **condensate**: heat pumps generate a considerable volume of condensate, particularly in winter months, which must be properly addressed;
(d) **noise**: heat pumps are noise emitters, so acoustics must be considered; and
(e) **insulation**: all water and refrigerant pipework must be properly insulated.

3.4.2 Air flows and access

Whilst ASHPs are versatile in their installation requirements, maintaining minimum clearances around them is essential to ensure correct operation. Clearances are required to achieve minimum air flows both on and off the coils to enable heat rejection. Inadequate clearance can lead to localized overheating, causing the heat pump to lock out.

Similarly to all large plant items, clearances are also required for cleaning and general periodic maintenance, where access is required to take apart and change internal components.

Clearance dimensions are included in manufacturers' information, along with guidance on placement to avoid interference from wind and general protection from weather. Strong winds can impede normal airflow and cause malfunction, whilst a build-up of snow could block the inlet/outlet, making it impossible for the heat pump to operate.

Enclosures are sometimes favoured around heat pumps to make them less conspicuous or to provide some acoustic damping. If desired, it is essential that manufacturer guidance on minimum clearance is heeded, so that airflow is not disrupted. Ground and water source heat pumps do not require clearance for air flow, but units will still require access for maintenance.

In the case of closed loop systems, provisions must also be made to access pipework junctions and valves, such that sections of a collector or borehole can be isolated should a leak occur. This is essential where antifreeze is used, as these chemicals can be toxic and limiting a leak to a localized area is a good first step in preventing spread.

For open loop systems, pumps and filtration equipment must be readily accessible so that strainers can be cleaned. If filters are allowed to clog with silt or vegetation, this could lead to a pump ceasing to operate and the heat pump system left without source media to draw heat from. Access to carry out essential maintenance is therefore essential.

3.4.3 Condensate

When an ASHP is working to provide heating the evaporator will be cold, causing moisture to condense out of the air.

Up to 8 litres of condensate can be produced by a domestic heat pump every hour. Condensate water is pure water and not acidic as it is produced with conventional gas boilers. The options for gravity drainage or pumped drainage into the local drainage system or a soakaway are acceptable.

3.4.4 Noise/vibration

The heat pump compressor generates noise and this can be an issue if the unit is installed in a residential area. Noise data is provided in manufacturers' literature and consideration can be made to the impact of noise. If noise is an issue, then low noise models or acoustic enclosures can be considered (noting the minimum air restrictions required).

Whilst in most cases vibration has been considered within the unit design, in certain cases vibration of the unit can be an issue. Potential solutions include flexible connections and anti-vibration mountings or pads.

3.4.5 Insulation

All refrigerant pipework must be insulated with a suitable grade of thermal insulation (see Figure 3.13). Failure to insulate the pipework will allow heat loss and reduce the performance of the heat pump. Pipework can also be prone to condensation formation that can lead to corrosion, so insulation must also be applied to prevent condensation.

Insulation should be specified to resist the maximum heat losses stated in BS 5422 *Method for specifying thermal insulating materials for pipes, tanks, vessels, ductwork and equipment operating*

Section 3 – Heat pumps

within the temperature range –40 °C to +700 °C. It is common for manufacturers and specifiers to advise the maximum permitted heat loss from which insulation thickness can be derived.

The thickness of pipework insulation will be based upon the conductivity of the insulation material and therefore it is not appropriate to simply specify an insulation thickness.

Figure 3.13   Heat pumps installation showing insulated pipework

3.4.6   Operating resilience

Any heat pump system must suit its application, and resilience measures should be dependent on its application. Where a continuous supply of heat is operation critical (for example, in a care home), it would be usual to provide several layers of resilience measures (perhaps twice as many heat pumps as were needed, known as '2n') to ensure that its purpose can be reliably fulfilled.

In domestic properties, whilst it is inconvenient to lose a supply of heat for a short time, this would not render it entirely uninhabitable and it would not be deemed necessary to install a second system in case the first were to fail. When conditions become extreme and outside of the design parameters, such as an extremely low temperature (typically –20 °C for most manufacturers) an alternative measure could be appropriate. This may apply in a climate zone that is known to regularly reach temperatures lower than –10 °C for prolonged periods.

In this circumstance, heat pumps can feature auxiliary resistance heaters, immersion coils in thermal stores or even backup gas-fired burners, either to provide continuous heat during defrost cycles or to temporarily replace a heat pump as the main heating generator during sub-optimal operating conditions. Certain product packages can provide this hybrid solution, and can also monitor gas and electricity tariffs to maximize financial economy.

In systems designed with more than one heat pump (for example, a cascaded or district scale installation), it would be usual to install additional heat pumps to enable continuous operation.

GSHPs and WSHPs are inherently more resilient than ASHPs, but can still encounter issues. The main threats to continuous performance can be localized freezing, if too much heat is extracted from the source medium, or leaks in the primary circulation system causing the heat pump system to fail. These issues occur due to erroneous design or installation, for example, if a collector installed in a body of water is too large, a ground loop is not installed deep enough in the ground or boreholes are drilled too close together. It is advisable to consult with specialists and follow guidelines set out by the Ground Source Heat Pump Association.

Beyond this, preventative efforts can be made to mitigate against leaks on closed loop systems by:

(a) installing boreholes with individual flow and return pipework to an accessible manifold, so that they can be individually isolated, and a single leaking borehole does not render the whole primary loop inoperable;

(b) ensuring there is provision for pressure monitoring to identify any leaks;

(c) not implementing automatic fill systems because automatically refilling can compensate for a leak, meaning it would go undetected; and

(d) monitoring the dilution of antifreeze to ensure adequate freeze protection is provided.

3.5 Summary

(a) Heat pumps are an efficient and reliable form of electric heat and are well suited to the UK climate.

(b) ASHPs are the most common solution for dwellings, but if it is affordable and possible opt for GSHPs as they are more efficient.

(c) Heat pumps use refrigerants which are harmful to the environment if they leak. Plan for monobloc solutions that seal the refrigerant inside the outdoor unit.

(d) Heat pumps operate most efficiently at low flow temperatures of 35–55 °C, which is lower than the 55–80 °C that traditional gas boilers operate at. Select complementary heat emitters that operate at lower temperatures.

(e) Noise levels, required airflows, the ability for automatic defrost, appearance and location should all be considered when selecting a heat pump.

(f) Select the heat pump solution with the highest performance and assess its efficiency through a combination of measures, including CoP, SCoP, SFP, SSHEE and its energy label.

≡ Section 4

Hot water considerations

This Section discusses the background to household hot water in the UK. It reviews the demand for hot water in the home, how it can be reduced and the risks concerning hot water in UK homes. Options for low-carbon heating are also discussed, with a focus on electric heating solutions from either direct electric, heat pumps or other renewable energy options.

4.1 Introduction

4.1.1 Traditional water heating in homes

For the average UK household, water heat represents about 20 % of total heat demand and is therefore a sizeable contribution to household heat demand.

Before the introduction of natural gas and the widescale adoption of gas heating, water would usually have been heated by coal using a back boiler in an open fireplace. This would be connected via large pipes to a galvanized steel hot water tank in the airing cupboard with the water naturally circulated. Water supply and pressure would be from a header tank often located in the loft. Water could also be heated using an electric immersion heater.

With the installation of gas heating, the steel tank was replaced with a copper cylinder and the water heated via a copper coil which was connected to the heating system.

Developments in gas boiler technology resulted in the introduction of the combi gas boiler and by the late 1990s it became the gas boiler of choice for most households. This combi gas boiler heated hot water on demand and for households with limited space it allowed the hot water cylinder to be removed, increasing available cupboard space. Households with hot water cylinders have declined from about 20 million households in 2000, to about 10 million households in 2020.

For households that have retained a hot water cylinder, the trend has moved away from cylinders that use a header tank and towards unvented systems. These do not use a header tank and deliver hot water at mains pressure.

4.1.2 Hot water demands in the home

The average hot water energy consumption of UK homes is approximately 4 kWh per day, based on an average household of 2.4 occupants and approximate daily usage of 80 litres of hot water per household. Hot water is used in the home for several sanitary purposes and includes washing, cleaning, cooking and producing hot beverages. Hot water is traditionally provided at mains pressure and stored in a hot water storage tank. From the storage tank, pipework is routed to feed hot water outlets and appliances. Typical hot water outlets include:

(a) kitchen sink (often provided with taps and extendable cleaning hose);
(b) wash basin in toilet and bathroom;
(c) shower;
(d) bath;
(e) appliances including washing machine and dishwasher (but only for models that require piped hot water connections); and
(f) instant boiling water taps for hot beverages at 100 °C.

Section 4 – Hot water considerations

Hot water demands vary according to the number of people living in a home, the size of the home and how many bathrooms are within the home. Demands also vary according to the lifestyles of the occupants, such as how often baths are taken and if more than one bathroom is provided in a home, and whether additional baths, showers, taps and appliances are used simultaneously. All these factors will affect the size of the water supply, the required water pressure, the amount of hot water stored and the sizing of pipes.

Considering water pressure in the UK, the water suppliers' statutory service standard level of mains water pressure is 10 m/head (or one bar). This means there is enough force/pressure to push the water to a height of 10 m. In practice, water pressure can vary between 1–2 bar (10–20 m). If a higher water pressure is needed, which may be the case in luxury homes, a water tank and a pressure booster tank is fitted to serve outlets, such as power showers.

An important aspect in sizing hot water storage is the recovery time to reheat water. This will depend on the lifestyle of the occupants. Some will require fast recovery to reheat the water if many people are living in the house, while a slower recovery will be acceptable if there are only one or two occupants in a home. The faster the recovery time required, the greater the power supply needed to reheat the water. Conversely, the slower the reheat time the lower the power supply. While a gas combi boiler provides instantaneous hot water, an electric heated water system can take one minute per litre to heat hot water.

4.1.3   Hot water design standards

It is important when providing either a new hot water system or modifying an existing hot water system to comply with regulations and standards. The most prominent standards being Building Regulations Approved Document G *Sanitation, hot water safety and water efficiency*, The Water Supply (Water Quality) Regulations and BS 8558:2015 *Guide to the design, installation, testing and maintenance of services supplying water for domestic use within buildings and their curtilages. Complementary guidance to BS EN 806.*

Building Regulations Approved Document G provides guidance on the supply of water to a property, including water safety, hot water supply, sanitation and water efficiency, that is, an easily accessible water supply that does not incur wastage. The guidance given is focused on the supply of sanitary, drinkable water to buildings, including cold drinking water, hot water supplied to bathroom appliances including baths, showers and sinks next to toilets (or in an adjoining room) and to food preparation areas in kitchens – and any water needed for the purposes of washing. Hot water systems also require extra safety precautions when they are unvented.

The Water Supply (Water Fittings) Regulations and Water Supply (Water Fittings) (Scotland) Byelaws play an important role in protecting public health, safeguarding water supplies and promoting the efficient use of water in premises across the UK. They set legal requirements for the design, installation, operation and maintenance of plumbing systems, water fittings and water-using appliances. They have a specific purpose to prevent misuse, waste, undue consumption or erroneous measurement of water and, most importantly, to prevent contamination of drinking water.

Additionally, BS 8558:2015 is the UK standard for the design, installation, testing and maintenance of services supplying water for domestic use within buildings and their curtilages. It provides best-practice recommendations on the design, installation, alteration, testing and maintenance of cold water and hot water supply systems for domestic use within buildings. The standard covers pipe work systems, pipe fittings and connected appliances installed to supply any building, whether domestic or not, with water for drinking, culinary, domestic laundry, ablutionary, cleaning and sanitary purposes. BS 8558:2015 only deals with low temperature systems, so it excludes systems designed to work with steam or high temperature water.

These regulations play an important role in ensuring that hot water systems do not foul or pollute drinking water supplies, and that they are efficient and operate safely. The regulations place requirements for devices, such as back flow restriction, expansion vessels, safety valves, isolation valves, pressure-reducing valves, thermal insulation and operational and maintenance requirements.

An important aspect of UK regulations controls the temperature of stored hot water that is required to reach a sterilization temperature of 60 °C to kill Legionella. Legionella bacteria cause Legionnaires' disease, which is a potentially fatal type of pneumonia, contracted by inhaling airborne water droplets containing viable Legionella bacteria, for example, from spray taps and showers. Although the Health and Safety Executive consider the risks from hot systems as low risk (due to the regular water flow and tank refill), care is still required to negate any risks. Hot water cylinders should regularly purge the water at 60 °C. To prevent scalding, hot water should be blended with cold water to reach 43 °C – this is usually achieved by using blending or mixing valves that mix hot and cold water.

4.1.4 Opportunities to reduce demand

Reducing the volume of DHW used in showers, taps and appliances will not only reduce the amount of hot water drawn but also the energy used to heat the water. Modern sanitaryware specifications such as low flow fixtures and shower heads offer choice for low water flow rates to conserve water and energy, and these can be specified or retrofitted in homes.

Other opportunities include timed flow restrictors or more commonly automatic taps used in wash hand basins to ensure that waste is minimized (see Figure 4.1).

Figure 4.1 An automatic bathroom tap to restrict water usage and waste

Modern energy efficient appliances such as dishwashers and washing machines will also provide low flow options with high energy ratings of A++ to A+++.

Therefore, when installing a new electrified heat solution, consideration should be given to the type of taps and appliances used and whether there is an opportunity to also conserve water and energy.

Another opportunity to consider is waste-water heat recovery. A waste-water heat recovery system can recover approximately 15 °C to water being fed into the hot water systems. There are two basic types of waste-water heat recovery device:

Section 4 – Hot water considerations

1. **linear drain heat exchanger**. The heat exchanger is located horizontally in the shower drain tray, so this option is much easier to retrofit. Up to 35 % of heat can be recovered depending on flow rate. The recovered hot water can then be utilized by being returned with a mixing valve, returned to the electric shower unit or returned to the heat source.
2. **vertical drain heat exchanger**. The heat exchanger encircles the drainpipe (not particularly suitable as a retrofit). These are more efficient and, although dependent on flow rate, can recover 60–70 % of the heat.

Figure 4.2 shows how hot water heat recovery works within a ShowerSave unit.

Figure 4.2   Hot water heat recovery in a ShowerSave unit

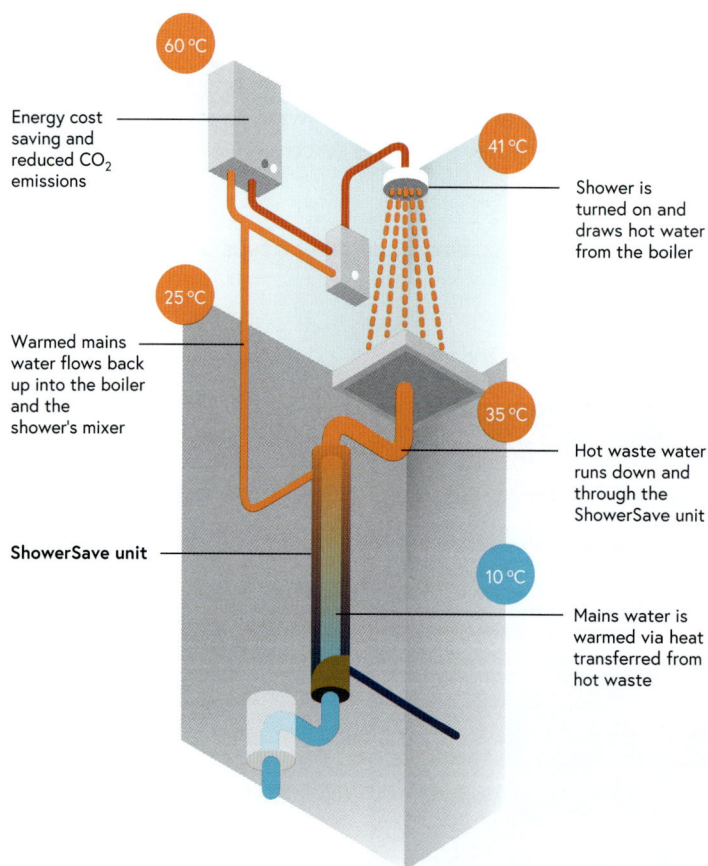

60 °C

Energy cost saving and reduced CO_2 emissions

41 °C

Shower is turned on and draws hot water from the boiler

25 °C

Warmed mains water flows back up into the boiler and the shower's mixer

35 °C

Hot waste water runs down and through the ShowerSave unit

ShowerSave unit

10 °C

Mains water is warmed via heat transferred from hot waste

4.2   Direct electric hot water production

As the trend moves away from water heating with gas combi boilers, there is a role for the use of direct electric water heating. This can either be to support heat pump solutions or to be used in applications where water demands are low, infrequent, remote or the physical installation of heat pumps is not practical as a standalone localized solution.

Options include central hot water cylinders, electric showers, local or point-of-use water heaters and hot beverage water heaters (see Figure 4.3 and Table 4.1).

Section 4 – Hot water considerations

Figure 4.3   Electric shower (left), hot beverage heater (middle) and point-of-use heater (right)

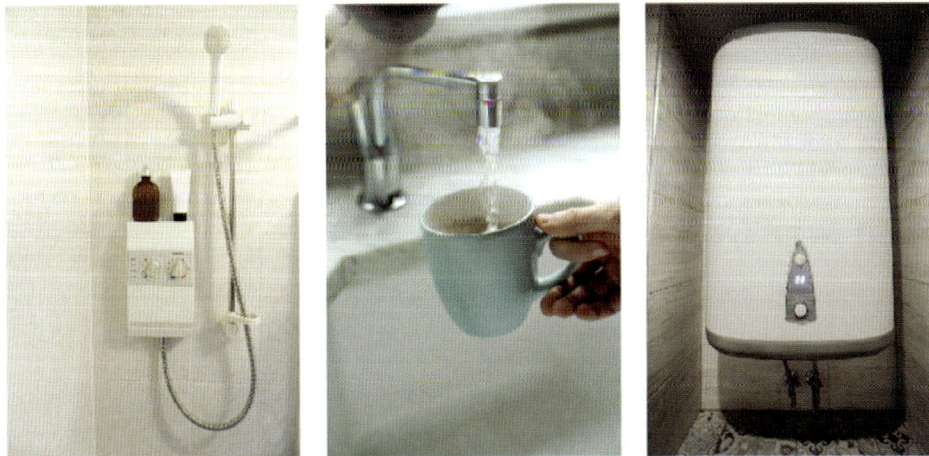

Table 4.1   Direct electric hot water options

Type	Applications	Typical power supply
Central water storage cylinder	These can store between 70–315 litres of hot water.	1 ×3 kW to 3 ×3 kW for a 9 kW immersion heater.
Localized electric shower	These do not store water and heat water instantaneously.	They are typically provided with a 7.5–8.5 kW direct electric heater.
Localized water heater for grouped outlets	Medium volume water heaters store higher volumes of hot water between 30–100 litres. These types of water heaters are suitable for multiple outlets including bathrooms.	Typically fitted with a 3 kW direct electric heater.
Localized point-of-use water heater	Localized water heaters for hand washing. These typically store between 6–15 litres of hot water for single outlets.	Typically provided with a 1–1.5 kW direct electric heater.
Localized hot beverage water heater	Hot drink making single kitchen mixing taps generating hot water at 98 °C with steaming hot water from an under-sink boiling tap.	Typically provided with a 1–2 kW direct electric heater.

There are several efficiency benefits with localized hot water heaters in that:

(a) there is no storage or long pipe run losses;
(b) only the water required is raised to temperature; and
(c) the water is only heated to the temperature required.

4.3   Heat pump hot water production

The majority of heat pumps providing space heating can also be used to provide hot water, both in new build and retrofit applications. However, system design should consider occupancy as heat pumps will switch from hot water to space heating, so it is best to produce a large store of water in the early hours for use during the day.

Section 4 – Hot water considerations

4.3.1 High temperature heat pump

High temperature heat pumps can generate up to 70 °C heat in domestic applications. These systems typically utilize hot water stratification storage tanks with the DHW being drawn from the top of the storage unit. When hot water is required, the heat pump operates at a higher output temperature, but lower CoP. It reverts to a higher CoP when space heating is required.

4.3.2 Cascade heat pump system

A cascade heat pump system (see Figure 4.4) is where two or more heat pumps operate from a single controller. The cascade operation enables stepped output to meet minimum and maximum loads. When a high demand for hot water is required, both heat pumps will operate together to meet the demand. Once the demand for hot water is achieved, the cascade mode switches back to heating mode operating at a reduced output (perhaps only one heat pump) or, in the case of summertime, switches off.

Figure 4.4 Cascade heat pump solution comprising two heat pumps

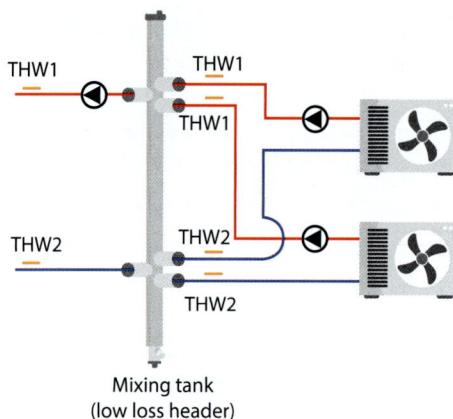

4.3.3 Hybrid heat pump system

A hybrid heat pump system is where a supplementary heat source is hydraulically interconnected into the heat pump system acting as a supplementary heat source. This enables the heat pump to operate at an efficient CoP. The heat pump provides heat into a thermal store or cylinder at around 50 °C. The supplementary heat source (for example, gas boiler or electric water heater) is used in tandem with the heat pump to increase the water temperature to 60 °C.

Depending on hot water demand, both the heat pump and supplementary heat source can work separately or in tandem as required (see Figure 4.5).

Section 4 – Hot water considerations

Figure 4.5 Hybrid heat pump solution using heat pump and gas boiler or electric water heater

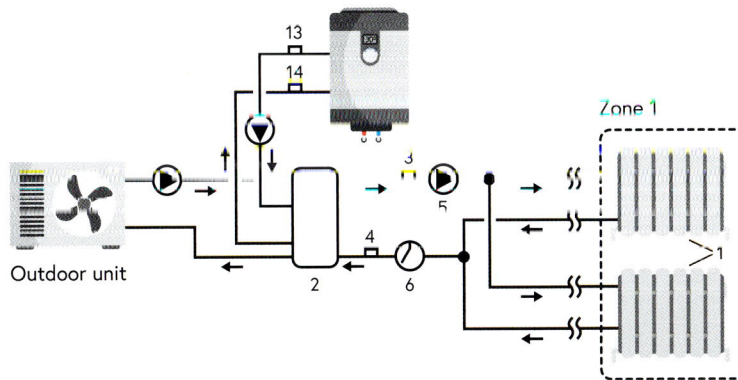

4.3.4 Hot water heat pumps

Another form of hybrid solution is the pairing of an electric space heating system (typically direct-acting panel heaters) with an ASHP that delivers domestic hot water – commonly referred to as a hot water heat pump (HWHP). These HWHPs consist of a ducted system to supply external air to an internally installed ASHP mounted on the top of a water cylinder. The heat pump collects energy from the air and converts this into heat, which is then transferred to the water within the cylinder – creating a store of hot water up to 60 °C for use in the dwelling. This technology allows for specification flexibility in both new buildings and retrofit scenarios where a more traditional heat pump that delivers space and water heating is not spatially or financially practical. These HWHP systems are recognized within SAP and can bring compliance and EPC improvements to projects when considered alongside the correct insulation and space heating measures.

Direct electric space heating solutions such as panel and storage heaters can benefit from being paired with HWHPs, in place of a more traditional technology such as a direct-acting water cylinder. Moving to a renewable hot water solution in this way brings system improvements in running cost and carbon emissions without major installation disruption. Application of electric hybrid technology combinations such as a HWHP and electric space heating can be a good compromise for properties where electrification is not easily achieved via the application of a whole-house solution heat pump technology.

4.4 Renewable energy hot water production

4.4.1 Solar thermal hot water system

A solar thermal hot water heating system, or solar thermal system, uses free heat from the sun to warm DHW.

The benefits of solar water heating include:

(a) the ability to generate hot water throughout the year;
(b) its utilization of a renewable form of heat; and
(c) its ability to be easily integrated into both a wet heating and hot water system.

Section 4 – Hot water considerations

A well-designed system can expect to collect useful energy greater than 450 kWh/m^2/year. However, performance is dependent on the location, orientation and panel-mounting angle of the collector.

Solar thermal hot water heating uses bespoke solar panels (collectors) that transfer solar energy to heat water that is then stored in a hot water cylinder and drawn off when required.

A supplementary heat source, such as a heat pump or electric immersion heater, is usually needed as a backup to 'top up' the tank due to variability in collector location and orientation that impacts the level of heat collected.

The two common types of collector are:

1. **flat plate**. The flat plate panel is typically roof mounted and often integrated into the roof structure itself. The efficiency of the panel is low and water temperatures returned to the DHW storage tank are limited and heavily reliant on the availability of direct sunlight.
2. **evacuated tube**. Constructed from a series of evacuated glass tubes and typically roof mounted, they are significantly more efficient than flat plate but are also more expensive. These are the most efficient panel type (conversion efficiency of 90 %) but also the most expensive. For a domestic property a typical evacuated tube solar collector system will cost about £3,000–£5,000 and will typically produce about 1,000–2,500 kWh of useful heat – or about 50 % of the building's hot water requirements.

Both systems are shown in Figure 4.6 and require a solar calorifier to enable the hot water generated to be stored until it is required. This also allows for other inputs into the hot water system including an electric immersion, heat pump or both.

Figure 4.6 Solar thermal hot water system components

4.4.2 Heat batteries

Heat batteries use a phase change material (PCM) to efficiently store heat and can utilize a range of thermal inputs, for example, immersion heater, combi boiler, solar thermal or heat pump. The unit replaces typical DHW cylinder or calorifiers, for example, a 3.5 kW unit would replace a 70-litre cylinder, ranging up to a 14 kWh unit replacing a 280-litre cylinder.

The PCMs used are non-toxic, non-flammable and capable of 50 years of use without performance degradation. For example, the Sunamp unit (see Figure 4.7) has been tested to over 40,000 charge and discharge cycles.

Section 4 – Hot water considerations

The units use a high flow rate heat exchanger delivering up to 24 litres/min, equating to 30 kW of instant heating. The units mix the water to the desired temperature reducing scald risk and there is no Legionella risk as no hot water is stored by the unit.

4.5   Summary

(a) DHW generation carries a risk from Legionella bacteria and backflow contamination. DHW design is well regulated to minimize this risk. Ensure all related standard requirements are followed.

(b) Reduce hot water demands by replacing water-using appliances with low flow taps and shower heads. This will reduce the size and capacity requirements of the hot water heating system.

(c) Where possible, recover heat from used hot water and recycle this energy using suitable and approved systems.

(d) Localized direct electric water heating can be an effective solution in smaller dwellings.

(e) Heat pumps can adequately provide DHW in dwellings, for example, high-temperature heat pumps, cascade and hybrid solutions, and is the most efficient method of generating hot water. Always include a backup electric immersion heater to add system resilience.

(f) Consider supporting renewable options to generate hot water such as solar thermal systems and heat batteries.

Section 5

Selecting electrified heat systems

This Section discusses the key electrified heating systems and key components available to the domestic market today. The main safety regulations that apply to safely install and operate these technologies will also be covered.

5.1 Introduction

5.1.1 Types of electrified heat

In general, there are two options for electrified heat systems: a heat pump-based system or a direct electric system. The changes in the Building Regulations Part L and the future homes consultation will favour those technologies that make the best use of the primary energy factors and that minimize running costs.

As a rule of thumb these technologies tend to favour certain types of dwelling. The current trend in the UK is for direct electric systems that suit smaller properties, such as apartments, where a heat pump application may not be suitable due to space limitations. The majority of heat pumps that have been installed in the UK are monobloc systems, which have the refrigerant gases sealed in the unit, much like a domestic fridge.

5.1.2 Additional considerations

A key consideration when reviewing electric space heating systems is the accompanying water heating solution. In most instances this is a significant load and in new dwellings it is often the primary load. Its technology and efficiency are likely to be linked to the space heating system either through the tariff that they have available, the controls that could combine them, or in some instances the technology that could provide for them both with a single appliance.

Building insulation and airtightness should be improved where possible before reviewing a heating system, as the selection and sizing could be different for a 'fabric first' approach.

5.2 Regulations and standards

5.2.1 Electrical safety

All electrical installation work in a home, garden, conservatory or outbuilding must meet the current Building Regulations. Apart from some forms of minor work, all electrical work must either be reported to the local authority building control or be carried out by an electrician who is registered under Part P of the Building Regulations. By law, all homeowners and landlords must be able to prove that all electrical installation work meets Part P. Local authorities can force developers, homeowners or landlords to remove or alter any work that does not meet the Building Regulations.

All electrical work carried out on site while installing both heat pumps and direct electric systems needs to comply with Part P and the current edition of BS 7671 *The IET Electrical Wiring Regulations* and should be completed by a qualified individual.

Section 5 – Selecting electrified heat systems

5.2.2 F-gas regulations

An F-gas is a fluorinated gas. Its use is regulated to ensure minimized usage and to limit the potential impact of accidental release. All types of heat pumps fall under the UK's F-gas regulations to some level.

As monobloc heat pump systems are hermetically sealed there is no need for the installer to be F-gas qualified. In the rare instance, however, that the sealed circuit needs to be broken into, an F-gas qualified engineer would be required.

Some heat pumps are defined as split type, where refrigerant lines run from the outdoor unit to an indoor hydro box or cylinder. F-gas certification is required for any work carried out on the refrigerant aspect of the system.

Anyone who services equipment that contains F-gases, such as refrigeration and air conditioning systems, where it involves handling, removing or replacing refrigeration gases must be qualified and appear on the F-gas register.

For any system that is not hermetically sealed, F-gas qualifications are required to:

(a) install new systems;
(b) service and maintain systems;
(c) check for leaks;
(d) recover gases; or
(e) decommission and dispose of old systems.

5.2.3 Microgeneration Certification Scheme accreditation

The Microgeneration Certification Scheme (MCS) is an industry-led quality assurance scheme that demonstrates the quality and reliability of approved products and installation companies. MCS-certified installers give customers the confidence that both the installer and manufacturer perform to the industry-expected level of quality. MCS is now seen as the quality mark for all renewable technologies, such as heat pumps. So much so that all government incentive schemes ask for an MCS accredited installation to qualify for the financial benefits.

5.2.4 Planning permission

Since 1 December 2011, the installation of a domestic ASHP has been classed as a permitted development, and needs no application for planning permission, provided:

(a) the ASHP installation complies with the Microgeneration Certification Scheme Planning Standards (MCS 020) or equivalent standards;
(b) the volume of the ASHP's outdoor unit does not exceed 0.6 m^3;
(c) it is only the first installation of an ASHP;
(d) all parts of the ASHP are at least 1 m from the property boundary;
(e) if the property is on land within a conservation area or World Heritage site, the ASHP must not be installed on a wall or roof which fronts a highway or is nearer to any highway which bounds the property than any part of the building; and
(f) the ASHP is used solely for heating purposes.

These permitted development rights apply to the installation, alteration or replacement of an ASHP on a house or block of flats, or within the curtilage (garden or grounds) of a house or block of flats, including a

Section 5 – Selecting electrified heat systems

building within that curtilage. A block of flats must consist wholly of residential flats (and should not also contain commercial premises).

5.2.5 SAP

The SAP is the methodology used by the UK government to assess and compare the energy and environmental performance of dwellings. Its purpose is to provide accurate and reliable assessments of dwelling energy performances needed to underpin energy and environmental policy initiatives.

The SAP methodology is used when calculating compliance to Building Regulations. The methodology is also used to judge the overall performance of a building and to determine the all-important energy performance certificate (EPC).

5.3 Direct electric heating systems

5.3.1 Power supply

Direct electric heating systems typically range in load from 200 W up to 3 kW. These products often come pre-fitted with a supply cable of suitable heat resistance and loading capacity. However, some ranges are manufactured with a pre-fitted plug for supply using a standard three-pin plug socket.

Exceptions are some high heat retention storage heaters, underfloor electric heating systems and electric boilers. In these instances, higher loads may be required and dual supplies might be involved, where switching to an off-peak tariff is achieved by switching power to an off-peak circuit.

Information on the requirements of the power supply is found in the installation instructions, and it is the responsibility of the specifier/installer to use the relevant regulations and guidance to correctly select the power supply, cable, switching and safety gear.

5.3.2 Operating costs

Direct electric heating is 100 % efficient at the point of use. This means that every kWh that the heating system draws through the meter heats the space in which it is installed. The technology required for a product to accurately maintain a room temperature is relatively cheap, and in many instances is regulated to a minimum standard (for example, the European Commission's Ecodesign Directive (2005/32/EC) Lot 20). This means that all energy goes into heating a space and that the space is rarely over- or under-heated.

Cost factors can be split into:

(a) **external factors**: for example, the external temperature, which cannot be changed.
(b) **associated factors**: for example, U-values of insulation. These cannot be controlled by the heater specification, but could possibly be improved by the building services engineer. Improvements in thermal efficiency through fabric improvement are a good first step in reducing the running cost of any installed electric heating system.
(c) **direct factors**: this would include the temperature that the space is maintained at, the duration it is maintained, and the cost of the energy used to maintain it. Additional energy saving features would also apply, such as the ability to cease heating if an open window or door is detected.

Section 5 – Selecting electrified heat systems

To minimize running costs of a direct electric heating design, understanding the occupancy requirements is critical. If the space will be occupied for long periods of time, or require a high temperature, then a system that minimizes the unit energy cost, despite a potential capital cost increase, is advisable. An example of this might be specifying a high retention storage heater in an older apartment's living room, if that is where the occupants spend most of the day. In contrast, a modern, well-insulated apartment in the city where occupancy might be infrequent would not benefit so much from this capital outlay, and a panel heater specification with a seven-day remote control capability would be more appropriate.

For most existing dwellings, off-peak tariffs are the best choice for low running costs, as they use cheaper rate energy to provide heating and hot water. Where tariffs are available, off-peak energy can be up to half the cost of peak-rate energy, with cost comparison models demonstrating running cost savings of up to 47%.

In many instances, however, dwellings or spaces within them can benefit from the application of direct-acting heating. Whilst running costs will be higher compared to off-peak alternatives, when used in areas where very little heating is required, they can help balance a specification to meet capital cost, space or load capacity restrictions. This is particularly true when paired with a renewable source of DHW production, such as a HWHP, where the primary load can be decarbonized.

5.3.3 Carbon

The carbon associated with direct-acting electric heating systems mirrors that of the electricity supply grid, with the significant decarbonization that has taken place over the last 30 years now being recognized within SAP. The carbon factor (kg CO_2/kWh) is now lower than that of natural gas, however generation and transmission losses mean that the primary energy factor of electricity is higher than that of fossil fuel equivalents. Primary energy reflects how much raw fuel is used to generate a unit of final energy. This includes the power used to create, transform and transport the energy from its raw form to where it is used. Electrical generation and transmission losses account for its relatively low primary energy score, despite its efficiency at point-of-use.

With a continually decarbonizing electricity grid, direct-acting electricity technology will be a progressively low-carbon option, however there are reasons why these are typically not the best solution for several applications.

Total load, particularly peak load, is an important consideration. Until flexible grid services are more common, direct-acting heating will typically draw power when other building services are also in operation. These systems will also draw energy at a 1:1 ratio with the kW requirement of the space they are heating. This means that, where peak avoidance is an issue, thermal storage heating is often a better option as energy can be taken at cheaper, off-peak times. Where total supply is a significant size, heat pump technology can be used instead to reduce the direct energy requirement on the electrical network by generating the majority of energy locally.

Direct and storage electric space heating solutions can benefit in terms of running cost and carbon by the addition of a renewable solution for hot water production. Applying technology, such as a HWHP, are a good compromise for properties where, for a range of reasons, electrification is not suitably achieved via the application of an ASHP.

5.3.4 Direct-acting heaters

There are several different types of direct-acting electric heating technologies. Whilst there are variances in the way in which they heat spaces and their occupants, ultimately the efficiency of all

Section 5 – Selecting electrified heat systems

technologies in this category is the same. All direct-acting electric heating is 100 % efficient at the point-of-use. In the UK, this means they share a common SAP category and a responsiveness rating of 1.0. 'Electric radiator' is a general term used to describe all the technologies listed below (see Figure 5.1).

Figure 5.1   Dimplex Q-Rad electric radiator (Reproduced by permission of Glen Dimplex Heating & Ventilation)

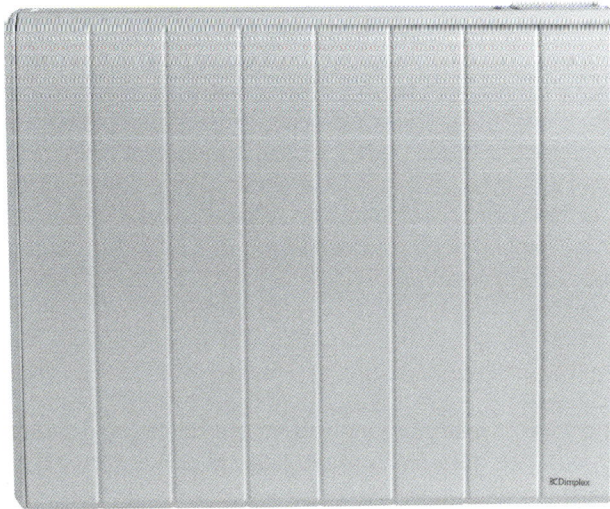

Panel heaters

A convective element inside a metal construction that heats a space via convection of the air within it (see Figure 5.2). Typically, heating will be noticeable in the first few minutes. Applications are broad, from bedrooms, conservatories and lesser-occupied spaces in older properties, through to a primary heating system in a more modern, well-insulated dwelling.

Figure 5.2   Dimplex PLXE Panel Heater showing controls (Reproduced by permission of Glen Dimplex Heating & Ventilation)

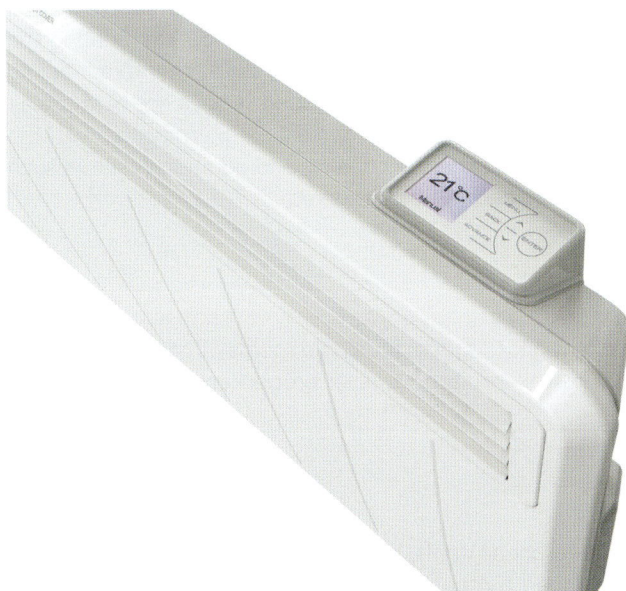

Section 5 – Selecting electrified heat systems

Convector heaters

Convector heaters are often used interchangeably with panel heaters. Typically, a convector heater is at the lower end of the specification range, comprising a simpler chassis and element construction, which results in a lower price.

Aluminium radiators

An aluminium radiator is a direct-acting heater (often filled with oil) that uses oil and/or aluminium to introduce inertia to the product. This has the effect of extending the duration of warm-up and cool-down of the heated space. This technology more closely mirrors the experience of a wet-system radiator in domestic applications, but there are no functional benefits of this inertia and running costs will be slightly higher due to the increased likelihood of delivering heat into unoccupied spaces. Heat will typically start to be noticeable in the first 15 minutes, depending on the inertial properties of the material used. These units are also heavier than other direct-acting heating.

Core heaters

A core heater is a direct-acting heater that uses a core of material to introduce inertia to the product. Similarly to the operation of an aluminium radiator, but with different core materials and therefore inertia properties. The material used (such as clay or ceramic) will slightly vary the warm-up and cool-down profile, with other attributes being consistent with other direct electric heating types.

Ancillary electric heaters

There are a number of ancillary direct electric heating technologies available, such as electric towel rails, bathroom 'downflow' fan heaters and tubular heaters. These technologies often complement other key technologies where there are unique environmental requirements to be met. Typically, these include a lack of space, a need for rapid, temporary output, or features not needed in other rooms, such as storing and drying towels.

Radiant heaters

Radiant heaters have a radiant element within the surface of the product which radiates heat and are used in external pub and restaurant smoking areas as well as stores and garages. Often described as 'heating the person, not the space', this technology is typically cheaper and physically shallower than other direct-acting heating. However, unlike most other electric space heaters discussed in this Section, it does not heat the air within a room. For this reason, radiant heaters have limited application in domestic environments, with the potential exception of hard-to-heat spaces where low insulation, low occupancy and additional factors such as high ceilings or use of a small area within a large room need to be considered.

Electric underfloor heating

Electric underfloor heating, which emits mainly radiant heat, is ideal for single rooms. It can, however, be the primary means of heat for the whole property in well-insulated smaller properties.

5.3.5   Electric off-peak storage heaters

Electric off-peak storage heaters store energy from a cheap tariff, often supplied at night. This stored heat is released into the space the next day, as required by the occupants. The technology is split into

Section 5 – Selecting electrified heat systems

traditional storage heating, and high heat retention storage heating (HHRSH), the latter of which meets greater insulation and control standards.

Off-peak heating is the preferred option for frequently occupied spaces, as operating costs are significantly lower than those of a direct-acting heater. This also allows for the specification of off-peak water heating, which provides DHW cost savings.

HHRSHs (see Figure 5.3) are recognized as offering additional efficiency over traditional storage heater technologies. In the UK, this results in a higher responsiveness rating and EPC improvement when using SAP. This is because the insulation is increased, meaning that the stored heat is delivered into the space dependent on occupancy requirements. Conversely, with historic storage heater technologies that have reduced insulation there are greater case emissions and the potential for heating an unoccupied space.

Figure 5.3 High heat retention storage heater

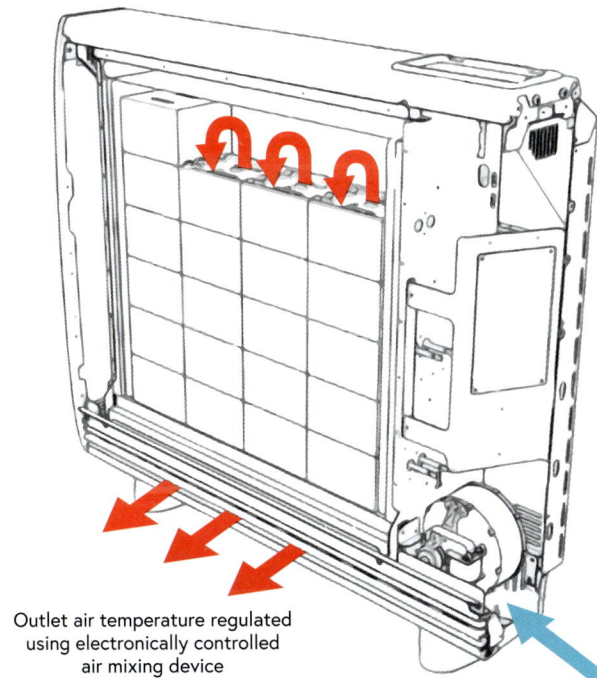

Outlet air temperature regulated using electronically controlled air mixing device

5.3.6 Electric boilers

Electric boilers use electricity to heat the water in a wet central heating system in a similar way to a gas or oil boiler. This energy can then be distributed into domestic spaces to heat them, or into a stored hot water tank for DHW use. Being suitable for wet central heating and DHW systems, an electric flow boiler can be used as a direct replacement for gas, oil, propane or solid fuel heating; or as separate extended systems, for example, to heat an extension where the existing primary heating system is close to capacity.

Electric boilers (see Figure 5.4), provide a versatile and safe way of heating water using electricity. They also provide a flexible installation as they do not require a flue or fuel tank storage like their fossil-fuel equivalents.

Section 5 – Selecting electrified heat systems

Figure 5.4 Redring Ascari direct electric water boiler (Reproduced by permission of Glen Dimplex Heating & Ventilation)

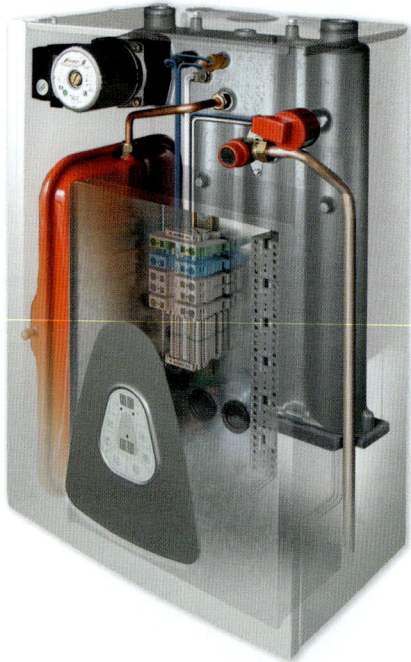

5.3.7 Summary of direct electric heating applications

Table 5.1 offers a summary of the direct heating applications discussed in this Section.

5.4 Heat pump systems

5.4.1 Key considerations

The design of a heat pump system is critical to ensure optimum performance and expected efficiency. Heat loss calculations should be carried out using the improved method outlined in Microgeneration Installation Standard: MIS 3005, *Requirements for MCS contractors undertaking the supply, design, installation, set to work, commissioning and handover of microgeneration heat pump systems* (prepared by the MCS). This requires each room to be accurately sized, taking into consideration all building fabric, windows and doors.

Once a total heat loss has been calculated, the selection choices of type of plant, equipment, components, emitter type and materials to meet the specification and performance requirements can start. The key considerations are:

(a) design flow temperature to meet building demands;
(b) heat emitter type;
(c) acoustic and vibration provisions: these must be adhered to during the installation phase, which usually involves flexible connection pipes and anti-vibration standing blocks;
(d) adequate provision for movement of pipework due to thermal expansion and contraction, hydraulic pressures and building movement;
(e) adequate chemical cleaning, flushing, addition of strainers and water treatment regimes;
(f) routing and selection of pipework both internal and external to the building;
(g) adequate thermal insulation using the appropriate weather-resistant products for external pipework and through-wall installations;

Section 5 – Selecting electrified heat systems

Table 5.1   Direct electric heating applications

Technology	Capital cost	Running cost	Main heat type	Benefits	Applications	Considerations
Panel heaters	Low	Medium	Convected	Low capital cost. Simple operation. Versatile. Easy to install.	New well-insulated properties. Less frequently occupied spaces in existing dwellings. Areas with low temperature requirements, e.g. bedrooms.	Running costs, where high heating loads might be present.
Aluminium radiators	Medium	Medium	Convected	Simple operation. Output profile is similar to wet radiator system.	New well-insulated properties. Less frequently occupied spaces in existing dwellings. Areas with low temperature requirements, e.g. bedrooms.	Running costs, where high heating loads might be present. Occupancy patterns.
Core heaters (clay/ceramic)	Medium	Medium	Convected	Simple operation. Output profile is similar to wet radiator system.	New well-insulated properties. Less frequently occupied spaces in existing dwellings. Areas with low temperature requirements, e.g. bedrooms.	Running costs, where high heating loads might be present. Occupancy patterns.

Table 5.1 (cont.)

Technology	Capital cost	Running cost	Main heat type	Benefits	Applications	Considerations
HHRSH	High	Low	Fan assisted	Low running costs. High output capacity. Simple operation.	All areas of existing dwellings, specifically those with high occupancy patterns or greater heating load requirements.	Off-peak tariff and supply availability. Capital cost.
Radiant heaters	Low	Medium	Radiant	Low capital cost. Simple operation. Versatile. Easy to install.	Poorly insulated spaces. Large spaces that are partly occupied. Spaces with high ceilings.	Location, as occupants will need to be in close to the panel
Electric boilers	High	Med	Various	Similarities with popular gas boiler systems.	Existing wet system electrification.	Running costs, where high heating loads might be present. Electrical supply capacity.
Electric Underfloor Heating	High	Med	Radiant	Simple operation. Space saving. Uniform distribution of heat.	New well insulated properties. Less frequently occupied spaces in existing dwellings. Areas with low temperature requirements, e.g. bedrooms.	Running costs, where high heating loads might be present. Occupancy patterns. Ease of installation.

Section 5 – Selecting electrified heat systems

(h) automatic controls; and

(i) selection of all brackets, supports and fixings, and any other items required to adequately complete the works.

It retrofitting, consideration should be given to the suitability of:

(a) the existing pipe work;

(b) the heat emitters and whether they need upgrading/resizing;

(c) the power supply and breaker sizing and whether these are adequate;

(d) the buildings fabric and insulation and whether these require upgrading due to heat loss values; and

(e) efficiency calculations and whether a new heating system would be more efficient than the existing system and improve both its carbon footprint and cost-effectiveness.

5.4.2 Assessment list

The application of a heat pump has several considerations for a new project. A site survey should identify where the outdoor unit can be located most conveniently, with consideration for access, condensate removal, minimizing sound to neighbouring properties and adequate airflow around the unit. Heat pumps require storage of hot water and the DHW cylinder must be installed in a suitable place that meets the recommended installation footprint for servicing. Minimum required water volumes for each property type should be recognized in the contractor's designs to ensure enough hot water for the occupants.

5.5 Heat pump wet systems

5.5.1 Monobloc ASHP and thermal store

Most heat pumps installed in the UK are monobloc type systems, where all the refrigerant gases are hermetically sealed within the outdoor unit. The outdoor unit is either floor- or wall-mounted outside. Heating flow and return water pipework connections are made between the outdoor unit and an indoor thermal store or heating and hot water cylinder. From this store and/or cylinder, heating and hot water is piped around the building. The full specification for products will be detailed in the manufacturers' data sheet, examples shown in Figure 5.5.

Figure 5.5 Examples of ASHP and heating and hot water cylinder data sheets (Reproduced by permission of Mitsubishi)

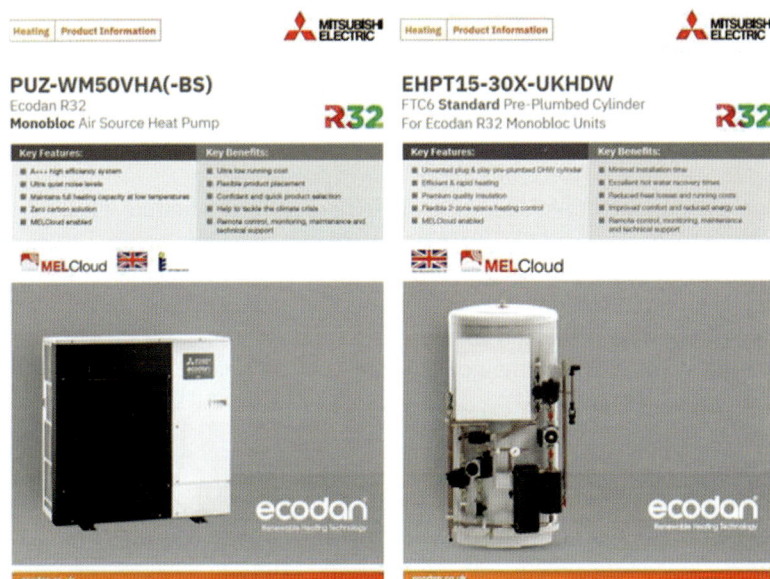

Section 5 – Selecting electrified heat systems

5.5.2 Emitter options

Heat pumps can be used with every type of emitter available on the market today, be that conventional or fan-assisted radiators, and are also suitable for use in underfloor heating (see Figure 5.6). Radiators need to be sized for the specific flow temperatures of the heat pump. Essentially, the lower the flow temperature the heat pump operates at, the more efficient and cost-effective the heat pump operation will be.

The key consideration is that the lower the design flow temperatures the more radiating surface area you need to emit the same amount of heat energy. The costs for the extra size in radiators are quickly regained. Extra wall space may not be required, if double panel double convectors are used.

Heat pumps work very efficiently with underfloor heating systems because these heating systems operate at very low temperatures and offer a large emitter surface. Underfloor heating systems with heat pumps usually comprise of zoned pipework that is connected to manifolds that can be individually controlled.

Figure 5.6   Underfloor heating installation with manifold prior to the laying of a screed floor finish

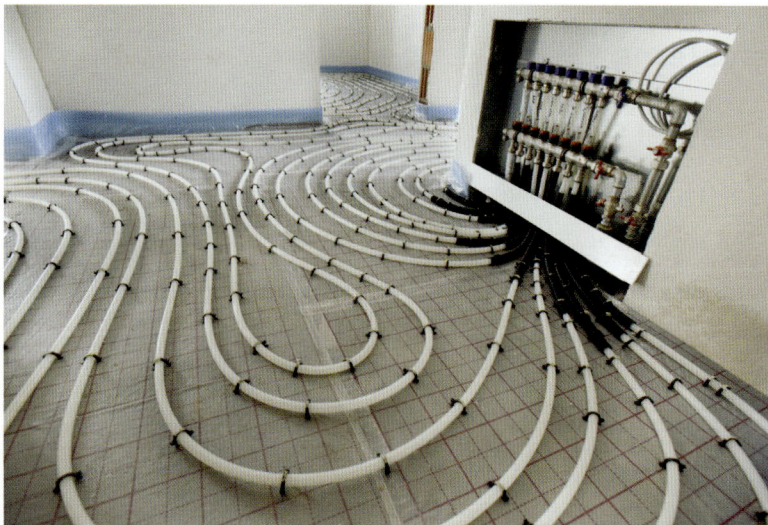

One increasingly popular solution is fan-assisted radiators (see Figure 5.7). These radiators feature small fans that aid air flow and allow the homeowner to quickly increase the temperature of a room if required. This can offer maximum efficiency at low water temperatures, while containing 15 % of the water normally stored within traditional radiators. The lower the water content, the quicker the radiator can heat up a room, allowing fast heat up times and efficient performance. Each of these radiators will require a low-voltage power supply to operate the fans. Most systems allow for the power supplies to be 'daisy-chained'.

5.5.3 Cooling capability

ASHPs are primarily used to provide heat into a property, but they can also provide cooling. This function is utilized in countries with higher average yearly ambient temperatures than the UK, such as Italy and

Section 5 – Selecting electrified heat systems

Figure 5.7   Dimplex Smart Rad – Compact fan-assisted low temperature radiator as an alternative to oversized radiators and underfloor heating (Reproduced by permission of Glen Dimplex Heating & Ventilation)

Spain. Standard heat emitters such as natural convection radiators and underfloor heating circuits are not best suited to cold water flowing through them as condensation will form when the warmer air in the room contacts the surface area of the heat emitter.

However, fan convection radiators with a built-in drip tray to catch the condensation below the heat exchanger, can be used for this purpose. Additional measures such as insulating the pipe work throughout the property and removing condensation to a drainage point would also need to be applied. As the need for cooling in domestic dwellings is less common in the UK, most manufacturers do not take design liability for this function. The provision of cooling also causes the installation to consume more energy and cost more money to operate.

5.6   Selecting controls

Control capabilities vary at an appliance, zone and building level, with different applications requiring functionality that might be standardized across that product category, but could also be a separate hardware or software specification.

5.6.1   Zoning

Most electrified heating systems have output and temperature capabilities for each space or room etc., so zoning is often simpler than for other system types. Zoning by product area can usually be completed using the on-board controls, with centralized controls being used to collect individual appliances and/or spaces into zones with common heating profiles.

In the mid- and high-specification end of these products, this is often achieved with the addition of a central control unit, operating either as a unique mobile/wall-mounted control (see Figure 5.8) or via

Section 5 – Selecting electrified heat systems

app-based technology on existing devices. Wireless communication across zones is often a feature of mid- to high-specification products.

Electric boilers may operate a more traditional zoning system, depending on the emitters used.

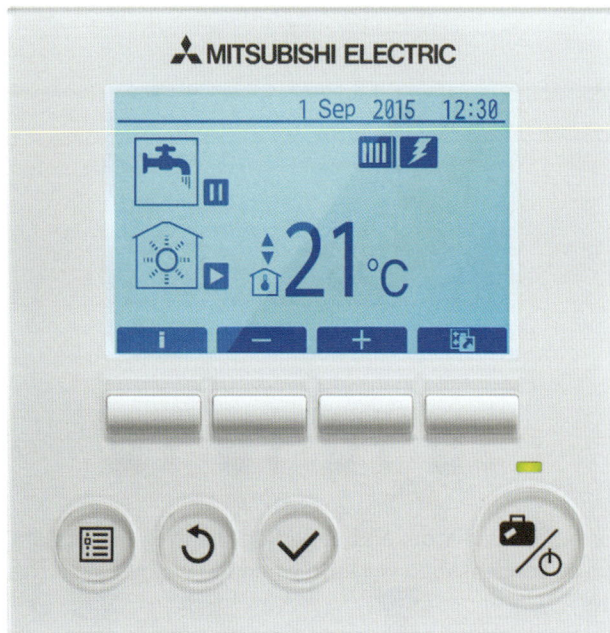

The capability to zone a system becomes more important as it grows in size and heating load, with substantial system efficiency benefits to be had when zoning larger, poorly insulated dwellings.

Modern heat pump systems should be capable of providing heating to two separate heating zones. Pre-plumbed and packaged cylinders are normally pre-wired and pre-plumbed with a three-port valve and circulation pumps from the factory as a one-zone heating and hot water system, but have the electrical connections and functionality for at least two heating zones.

Pre-plumbed cylinders are usually manufactured with a low loss header with blanked off connections for the second heating zone. They can be adapted to a two-zone heating system by connecting the flow and return pipes for the second heating zone to the low loss header, and with the addition of one extra circulation pump.

As both heating zones are drawing the heat from the low loss header, both heating zones will operate at the same flow temperature and cycle on/off on their own room thermostat demand. This would be ideal for controlling the temperature of a house with radiator circuits on both the ground floor and first floor.

Applications that require independent flow temperatures for each heating zone, require a mixing valve to blend the warmer flow temperature down to suit the required lower flow temperature of the second

Section 5 – Selecting electrified heat systems

heating zone. A good example of this kind of application would be a warmer radiator circuit on the first floor and lower flow temperature under floor heating circuit on the ground floor. The system will target the required flow temperature for the zone currently in demand and when both zones are in demand at the same time the system targets the higher temperature radiator circuit flow temperature and uses the mixing valve to reduce the water temperature to meet the underfloor heating circuit requirements. This type of application requires additional temperature sensors to monitor the flow and return temperatures for both heating zones (see Figure 5.9).

Figure 5.9   Heating zone layouts for one and two zones

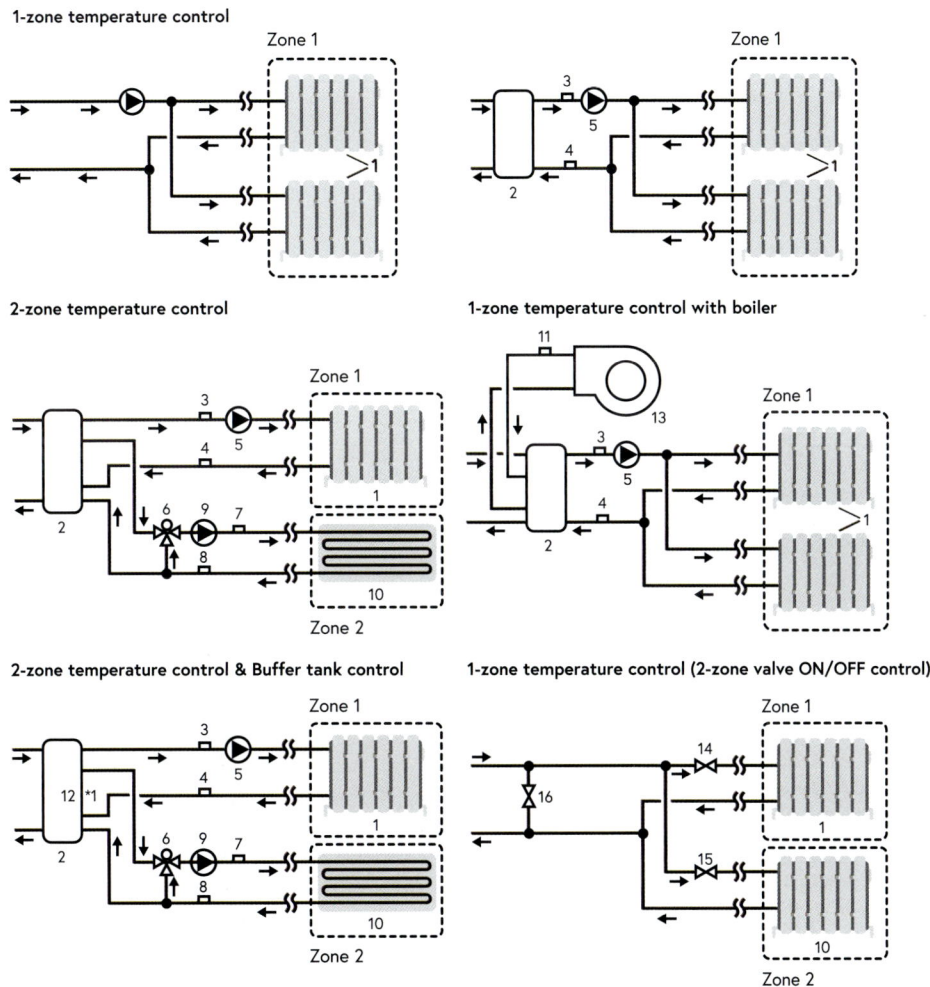

1-zone temperature control

2-zone temperature control

1-zone temperature control with boiler

2-zone temperature control & Buffer tank control

1-zone temperature control (2-zone valve ON/OFF control)

1. Zone 1 heat emiters (e.g. radiator, fan coil unit) (local supply)
2. Mixing tank (local supply)
3. Zone 1 flow water temp. thermistor (THW6)
4. Zone 1 return water temp. thermistor (THW7) ⎤ Optional part: PAC-TH011-E
5. Zone 1 water circulation pump (local supply) ⎦
6. Motorised mixing valve (local supply)
7. Zone 2 flow water temp. thermistor (THW8)
8. Zone 2 return water temp. thermistor (THW9) ⎤ Optional part: PAC-TH011-E
9. Zone 2 water circulation pump (local supply) ⎦
10. Zone 2 heat emitters (e.g. underfloor heating) (local supply)
11. Boiler flow water temp. thermistor (THWB1) ⎤ Optional part: PAC-TH012HT(L)-E
12. Mixing tank thermistor (THW10) *1 ⎦
13. Boiler (local supply)
14. Zone 1 2-way valve (local supply)
15. Zone 2 2-way valve (local supply)
16. Bypass valve (local supply)
*1 ONLY Buffer tank control (heating/cooling) applies to 'Smart grid ready'

Section 5 – Selecting electrified heat systems

5.6.2 Thermostats

In many instances, thermostats on direct electric heating devices are regulated. The European Commission's Ecodesign Directive (2005/32/EC) Lot 20 regulations require that an installed domestic electric space heater has 'seven-day electronic time and temperature functionality with electronic thermostat'.

This means that each appliance can accurately control the temperature in its space, and that the user can programme it over a seven-day period. In practice, the functionality and user experience of these thermostats can vary greatly, from basic, icon-based functionality to app-enabled smart thermostat functionality, such as geo-fencing, temperature prediction and automatic warm-up and cool-down energy saving functionality (see Figure 5.10).

Figure 5.10   Smart thermostats and controls

Because wet systems typically operate on one loop of emitters, localized room thermostatic control is not always available and often not accurate or electronically controlled. Typical specifications would include a single thermostat paired with the boiler (either fixed or mobile within the dwelling) to achieve the required temperature at that location. The next development would be zoned thermostats or individual emitter/space thermostats. These are likely to be TRVs if the electric boiler is paired to traditional radiator-style emitters. These systems can be smart and provide excellent levels of thermostatic control, but this is typically achieved through the specification of additional hardware to suit the application requirements.

5.6.3   Time control

The European Commission's Ecodesign Directive (2005/32/EC) Lot 20 regulations have placed a minimum standard of electronic time control functionality on all direct electric heaters used primarily for the purpose of heating populated spaces. This means that it will be standard for users to have some form of electronic on-board time control.

Improvements on this minimum standard include better user experience, larger screens, more timed periods, pre-set profiles and the ability to easily adjust settings, for example using boost, advance, 'leaving' or other modes to override the timer.

The time control for both heating and hot water in wet systems is operated through a main two-channel controller. Many state-of-the-art systems use web-enabled apps so there is no need for external two-channel time clocks. Either way, independent on/off timer schedules can be programmed for heating, hot water and the prevention of disease, for example, a daily or weekly temperature boost can be programmed which can prevent bacteria such as Legionella from growing. The schedules can be programmed from the main controller connected to the heat pump or through the app controls. Programmable room thermostats can be applied to the heating system if required, on bespoke applications.

5.6.4 Energy-saving functionality

There are a number of additional energy-saving functionality options.

Open window sensing technology

If a window (or door) was left open by accident, an additional heating load would be required. Without intelligence, appliances could just run harder, potentially at maximum output until the window or door were closed. With open window sensing technology this heat loss can be managed electronically using an algorithm to log the standard heat loss experienced within a space, noting anomalies and reducing or ceasing operation until parameters return to normal. This technology ensures the output from the heater is reduced when heat escapes from a room and avoids wasting both energy and money.

Predictive start controls

Higher-end products can offer functionality that facilitates a self-learning delayed start function. In a similar way to the open window sensing function, this 'learns' the thermal characteristics of a room and determines how long the appliance would need to operate to reach target temperature at a set time. This is based on factors, such as room size, unit output and heat losses, regardless of the prevailing weather. The heater will work out when it needs to start heating by maintaining a measurement for the heat up and cool down times of the room, so that it reaches the user-defined target temperature at a specified time.

This minimizes wasted energy and can deliver cost savings for users. In Figure 5.11, the occupant plans to get up at 7 a.m. and their desired temperature is 21°C. With a traditional system, the user would need to estimate when to set the heating to come on to ensure it preheats the room in time. Depending on the settings and the starting temperature, this could mean that the room is still cold when they enter it, or that it has been warm for some time before it needed to be.

With this functionality, an appliance can anticipate when it needs to switch on heating to reach 21°C by 7 a.m., preheating for shorter periods when the weather is mild and longer periods in winter.

State-of-the-art control systems now include intelligent room temperature control as standard, which is the most efficient way to run a heat pump system rather than, for example, a fixed flow temperature. This is complemented by system energy monitoring, which will show both consumed and produced energy.

Many systems are now being produced smart grid-ready and offer improved, easy system set up through a step-by-step installation wizard, which ensures systems are set up to match the demands of the home and the user.

Section 5 – Selecting electrified heat systems

Figure 5.11 Predictive start heating control example

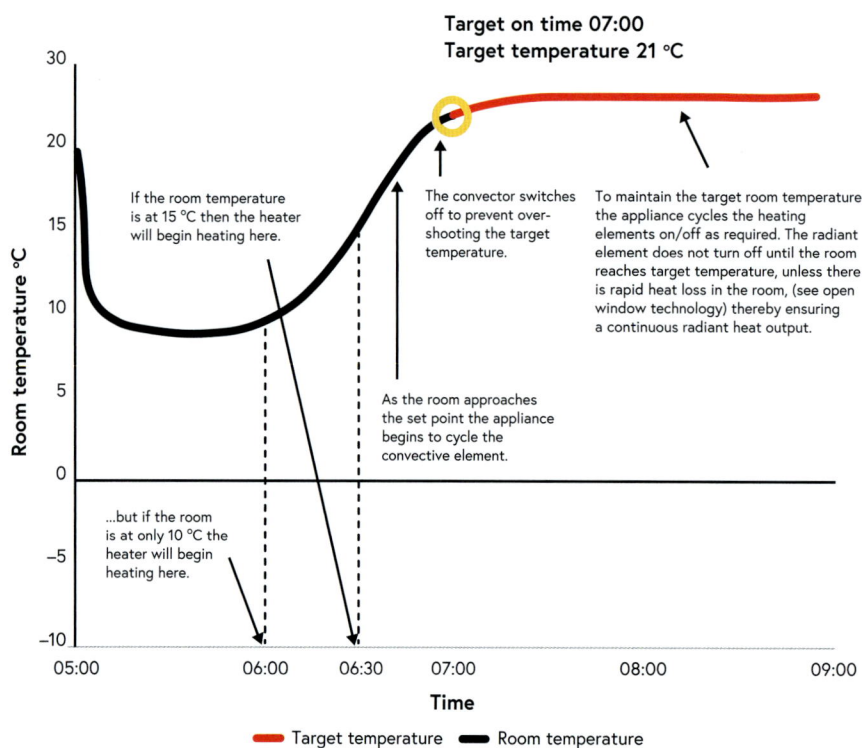

Target on time 07:00
Target temperature 21 °C

If the room temperature is at 15 °C then the heater will begin heating here.

The convector switches off to prevent over-shooting the target temperature.

To maintain the target room temperature the appliance cycles the heating elements on/off as required. The radiant element does not turn off until the room reaches target temperature, unless there is rapid heat loss in the room, (see open window technology) thereby ensuring a continuous radiant heat output.

As the room approaches the set point the appliance begins to cycle the convective element.

...but if the room is at only 10 °C the heater will begin heating here.

Room temperature °C

Time

— Target temperature — Room temperature

5.6.5 Non-user controls

There are a range of non-user controls to consider in the specification of direct-acting electric heating systems. Non-users could include facilities management, landlords, building owners and bill payers. Non-user controls include:

(a) **energy reporting**. Many direct electrical systems now have the capability to accurately estimate or measure energy usage. This allows for the interrogation of data to reduce energy costs, improve user comfort and address potential building issues, such as mould growth in an area that is not being heated.

(b) **system diagnostics**. Direct electric heating systems are well suited for diagnostic functions, where thermostats, fans, controls, elements and other electrical components can be responsively fitted with error codes that facilitate one-visit repairs with correct componentry. Proactively, they can also predict components nearing their end-of-life and facilitate planned maintenance without the inconvenience of down time.

5.7 Operation and maintenance

5.7.1 Operation

Operation of electrified heating systems is usually straightforward. There are few instances of users needing to operate mechanical controls and a high potential for automation, meaning the specification of high-quality electronic control systems with an intuitive user experience can be used to simplify operation.

Section 5 – Selecting electrified heat systems

A potential complication can be that electrified systems are suited for modular specification, and so legacy systems or varied new specifications can see several different control systems operating within one building. This can complicate the user experience and, therefore, packaged solutions are recommended where possible.

All systems are now provided with easy-to-use interfaces, many of which can be added to a building management system (BMS) or can operate from apps.

The most important aspect of operating an electrified heating system is to ensure that the heating system operates within the desired occupied time settings and to an energy efficient temperature setting for both daytime and nighttime use. Typical settings would be during wintertime 21°C in daytime and 13°C overnight.

5.7.2 Maintenance

Many direct electric heating systems require little to no maintenance. Regular electrical safety checks of the installation are recommended (see Figure 5.12). However, in the case of electric boilers, direct-acting and electric storage heating technologies the only requirement for a serviceable product lifespan of 15 or more years is to occasionally vacuum grilles or openings to remove dust.

Figure 5.12   Engineer undertaking standard maintenance (Reproduced by permission of Mitsubishi)

Heat pump systems must be maintained and serviced throughout their working life, typically on an annual basis. Most manufacturers will require proof of service to comply with warranty conditions and also to meet the demands of the government's Renewable Heat Incentive (RHI) scheme, which requires an annual service as a minimum.

The annual heat pump service typically comprises:

(a) cleaning the outdoor unit, while paying particular attention to the evaporator coil where the heat exchange happens. This area often attracts moss and leaves and may block airflow.

(b) antifreeze checks to ensure minimum glycol levels are maintained. This is essential to ensure the outdoor unit heat exchanger is protected.

(c) cleaning of all water filters, removing excess debris to ensure filters are in perfect condition.

(d) removal of any trapped system air. Even with automatic bleed valves, air may get trapped in the system and can cause a drop in performance.

(e) checking system pressure, which is essential to ensure efficiency.

(f) checking flow rate and adjusting if necessary. Flow rate is critical to ensure efficient energy exchange.

(g) checking and optimizing controller settings. Feedback from the owner can help maximize the performance, for example, the compensation curve can be adjusted if original settings are not performing for optimum efficiency.

(h) checking the immersion heater function to ensure Legionella protection is activated.

(i) testing unvented safety equipment as detailed in Part G3 of Building Regulations.

(j) charging expansion vessel if required.

(k) testing heat-up performance in both heating and hot water mode.

(l) checking and repairing the insulation of pipework.

5.8 Summary

(a) Small apartments that are highly insulated are well suited for direct electric heating solutions.

(b) Heat pumps are suitable for both new dwellings and for well-insulated existing dwellings. The monobloc system with a thermal store offers a good solution in both cases.

(c) In new dwellings, opt for underfloor heating or low-temperature fan-assisted radiators to provide heating in rooms. In retrofit applications, carefully assess the temperature and output of the existing heat emitters and insulate then ventilate to achieve output. However, it may be necessary to replace existing heat emitters with suitably sized low-temperature fan-assisted radiators.

(d) Zoning and control should be carefully planned for the occupiers and their lifestyle. Smart solutions are now widely available and are the preferred choice.

(e) The installation of all heat pumps should be carried out by an accredited MCS approved contractor.

(f) Heat pumps require annual servicing and maintenance, and this should be planned at the time of installation.

■ Section 6

Electrical supply considerations

When transitioning to electrified heat the increase in electrical load demands must be fully understood along with the constraints at the point of supply. If not, there may be consequences, such as installation overloads, poor power quality and operational inefficiencies.

This Section discusses the requirements for electrified heat on electrical wiring systems, including existing supplies, new supplies and metering arrangements. It explores the circuit arrangements, circuit protection and specific wiring requirements. The options for smart design are included, with consideration for integrated thermal storage, electrical energy battery storage, photovoltaics and electric vehicle (EV) charging.

6.1 Electrified heat and supply capacity

6.1.1 Supply constraints, perceived load demands and myths

Concerns about the constraints of existing electrical supply infrastructure and the perceived increasing load demands has resulted in several myths. These can be addressed by comprehensive surveys, accurate record drawings and correct assessment of existing and planned loads.

Any energy transition for electrified heat, using decarbonized electricity as the pathway, will involve careful energy management at the point-of-use and improved infrastructure. It takes time, investment and an understanding that user behaviours need to adapt to reduce system inefficiencies.

Table 6.1 considers some of the myths around electrified heat and its impact on power supplies.

Table 6.1 Urban myths concerning electrified heat

Urban myths	Commentary
'The use of electrified heat puts a strain on the National Grid.'	Electrical loads are increasing but electrified heat is only part of the picture. EVs may also add to the concerns, as well as the increased use of air conditioning. The National Grid has projected load assessments that include all these loads.
'Existing supply cables may not be robust enough to cope with this increased demand, causing cables to overheat and operate inefficiently. Existing generation capacity may not be able to accept all of the demand at peak times.'	The distribution network operators (DNOs) are continually assessing new electrical supply applications and planning accordingly for infrastructure enhancements to cope with increasing demand. Where electrical supply capacities are indeed too small, applications can be made for larger supplies.
'With all the energy savings being made from light-emitting diode (LED) lighting replacements, increased electrified heat demand will not be a problem.'	Savings from newer technology, such as LED lighting, is only part of the picture and will provide limited help. Smarter controls of all electrical loads may provide better energy efficiency and energy saving results.
'I am more concerned about existing buildings' electrical infrastructures not being big enough to cope with the additional load of electrified heat.'	There might be a problem, but careful design and load management will help to mitigate this. Electrical installations will need more controls to limit when loads are used and to be able to switch off when not required. It is, however, possible that new supplies and infrastructure will be needed, requiring full electrical refurbishment and upgrade to existing buildings.

Section 6 – Electrical supply considerations

The reality is that the decarbonization of space heating and water heating will place additional demand on electricity supplies. Electrified heat is a flexible energy vector and provides a suitable channel from a range of diverse renewable energy suppliers. This flexibility also allows loads to be shifted away from peak times, reducing costs and demands on infrastructure. Improvements in building construction and fabric will also reduce demands and allow warmth to be retained for longer.

Whilst using renewables reduces the carbon overhead, there is a concern of the relative cost of electrical energy and rising energy costs. The predicted increases are due to levies to pay for renewable energy generation, network control and transmission and distribution charges.

The electrical load needed to power electrified heat will vary from building to building, according to its type, size, age and levels of insulation. In a domestic setting, the power required for heat pump electrified heat systems can vary between 1.5–9 kVA.

An energy efficient heat pump is likely to be the most energy intensive device in the home. This may be an issue in retrofitting applications where historically DNO assessments of large developments may limit supplies to a nominal maximum demand of only 2 kVA per home and all other circuits and appliances.

6.1.2 Planning for new domestic installations

There are extensive requirements for a modern domestic consumer unit. It may require connections to accommodate for:

(a) mains isolation and photovoltaic (PV) isolation;
(b) surge protection;
(c) PV microgeneration and metering;
(d) mechanical plant and services, including electrified heat;
(e) electrical energy storage;
(f) EV charge points; and
(g) typical domestic small power and lighting loads.

Sufficient modular devices and proprietary cable and busbar connections should be provided within the consumer unit to accommodate the required circuits.

Refer to BS 7671 regarding measures for circuit protective cables and equipotential bonding.

Table 6.2 describes a theoretical three-storey townhouse consumer unit and provides some examples of the final circuits that could be required. In this example, residual-current circuit-breakers (with overcurrent protection) (RCBO) are used where users can interface with equipment, such as sockets and lighting, or the services are external to the main building. Otherwise, miniature circuit breaker devices (MCB) are deployed for fixed appliances and other services. Using a split load board with RCD controls several circuits and could provide a single point of failure for large parts of the consumer unit.

New legislation on gas supplies to domestic boilers and the declared legislative intent on heat pumps from the UK government, changes the landscape for domestic electrical consumers units. Gas supplied to new domestic buildings boilers is likely to be phased out within approximately five years, and there will be a transition to electrified heat for space heating and hot water.

This transition will impact on a typical consumer unit for a new build in terms of the number of outgoing ways required. A larger consumer unit may be required (possibly including dual rail units) to provide enough room to accommodate all the protective devices.

Section 6 – Electrical supply considerations

Table 6.2   Example circuit data for incorporating electrified heat in a three-storey domestic consumer unit

Way	Module description	Location	Protective rating (A)	RCBO/MCB
A	Surge protection	Consumer unit		
B	PV isolator (interlock with mains isolator)	Consumer unit		
C	Main isolator	Consumer unit		
D	EV charge point	Garage	40 A	RCBO
E	ASHP	External	20 A	RCBO
F	Hydrobox	Ground	20 A	MCB
G	Backup heater (underfloor heating)	Ground	16 A	MCB
H	Booster heater (DHW)	Ground	16 A	MCB
I	Kitchen hob	Ground	20 A	MCB
J	Kitchen oven	Ground	32 A	MCB
K	Lighting: ground floor, garage, exterior	Ground	10 A	RCBO
L	Lighting: ground and first floor	Ground	10 A	RCBO
M	Sockets (ring final circuit): garage	Garage	32 A	RCBO
N	Sockets (ring final circuit): ground floor	Ground	32 A	RCBO
O	Sockets (ring final circuit): ground and first floor	Ground and first	32 A	RCBO
P	Sockets (ring final circuit): kitchen	Ground	32 A	RCBO
Q	Mechanical ventilation with heat recovery	First	16 A	MCB
R	Smoke and heat detectors	All floors	16 A	MCB

Historically, electrified heat was produced through storage heaters for space heating and twin element immersion heaters for hot water and operated on a dual tariff meter (Economy 7 off-peak with lower tariffs) with its own consumer unit of up to six ways. This would have been installed beside a normal consumer unit of up to eight ways for general use and metered separately from the storage heaters.

A modern approach would be to host all circuits in the same consumer unit. This may be supported by a smart meter to assist with monitoring and controls. The number of sub-circuits and protective devices might require a single-phase consumer unit with twin mounting rails holding up to 16 protective device modules each.

Electrified heat will require more than one circuit (see items E, F, G and H in Table 6.2). All circuits will be radials and heat pumps will usually have backup heaters to consider. This equates to a traditional gas boiler system having a hot water storage tank with an integral electrical immersion heater.

Figure 6.1 illustrates the electrical supplies needed for a heat pump solution in a domestic property. Note the need for thermal hot water and heating storage and controls wiring.

Section 6 – Electrical supply considerations

Figure 6.1 Electrical and thermal requirements for heat pumps

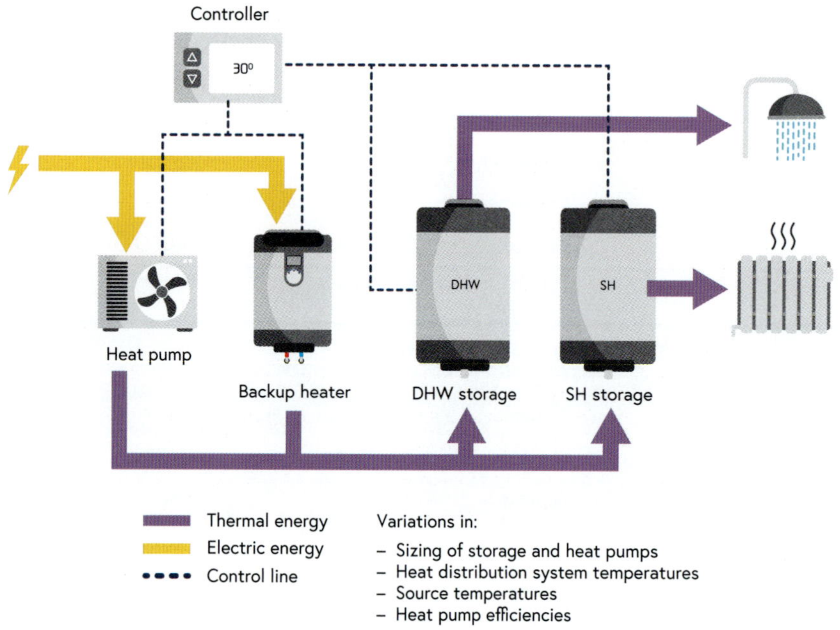

6.1.3 Planning for retrofit domestic installations

Providing electrified heat in an existing individual domestic installation can be a challenge. There are a number of additional circuits required for a heat pump installation and a careful comparison is needed to check the options available in the existing consumer unit.

A gas-fueled installation, with a standby electrical immersion heater, will typically use two protective devices, one for the primary system gas boiler control circuit and pumps and another one for the secondary immersion heater system.

A replacement heat pump installation for both space heating and hot water could use four protective devices. There will be a need for primary system controls and pumps, and primary supply for the heat pump. Both the space heating and the hot water will need separate secondary heating facilities. This all imposes a larger load on the installation intake, which could be twice the previous load.

A full assessment must be carried out on the existing installation before a heat pump is added to the existing consumer unit. A qualified electrical contractor should fully assess the existing consumer unit and supply first. This includes:

(a) rating the existing supply, bearing in mind that the main intake fuse is sealed;
(b) checking the condition of the existing supply cable;
(c) assessing the number of spare ways; and
(d) assessing the number and ratings of the existing protective devices associated with existing heating and hot water.

Upgrade strategies may include:

(a) a replacement electrical supply – either a larger single-phase supply or possibly a domestic three-phase supply;
(b) a replacement consumer unit;

(c) a sub-distribution consumer unit specifically for the electrified heat circuits;

(d) replacement circuits connected to the existing consumer unit; and

(e) the addition of new circuits to the existing consumer unit for the electrified heat systems.

If an existing property is undergoing a full-scale refurbishment, as part of a sustainability drive, electrified heat may be just one part of the project. PV, energy storage and EV charging should also be considered. In these instances, consider a new supply to the building and installing an upgraded consumer unit, rather than adapting the existing installation.

6.1.4 Power supplies, load diversity and maximum demand

A typical domestic single-phase (230 V) installation, for a house, may have an 80 A supply fuse which can support a peak load of around 18 kVA. Apartments may have smaller supplies.

Existing loads in peak periods might include electric cooking (up to 6 kVA), electric showers (up to 8 kVA) as well as lighting circuits and socket circuits for small appliances (for example, 2.5 kVA for kettles). The addition of heat pumps (for example, 7.4 kVA and rated at 32 A) could easily overload the intake if all appliances were operating at their maximum at the same time.

It is necessary to assess all the loads when calculating the full electrical requirements for a property. A simple summation of all loads will exceed the intake fuse rating. However, it is unlikely that all these loads will be operating simultaneously and therefore a load diversity factor is used to determine the probable maximum demand.

Historically, this maximum demand figure is a best guess based on agreed probable reductions in the load, for example, not all lights being on at any one time, not all sockets being on and not all heating loads will be heating spaces or water cylinders.

Diversity factors and maximum demand are mentioned in BS 7671 and developed further in the associated guidance notes. The current published diversity factor values were developed decades ago and well before the development of modern electrical loads such as heat pumps and EVs.

DNOs are still evolving their own response to increasing demands on the grid networks. Some current guidance is based on integration of EVs across larger housing estates. However, DNOs do not provide guidance on individual installations.

For modern domestic installations, controls linked to smart meters will be advantageous, and all electrified heating loads will have basic timer and thermostatic controls.

Adding an integrated control system can ensure that high energy use circuit loads either do not work simultaneously or operate at reduced output during peak times. For instance, if an EV is plugged in for ad hoc charging and then the heat pump switches at a pre-programmed time, the controls could reduce the energy provided to the EV so that it continues to charge at a slower rate and the heat pump operates satisfactorily.

Table 6.3 illustrates an estimation of the total demand of individual loads and their operational hours during different 6-hour periods. This gives an indication of periodic demand on cycles through a 24-hour day. Loads can be programmed to connect during off peak periods, such as for EV charging, and when electrified heat or lighting is required. Energy storage, such as thermal storage for heating and hot water, could be connected in off-peak hours too if programmed correctly.

The amount of current required at single-phase and three-phase voltages is shown in Table 6.3. Increasing demands on domestic installations will require more consideration of three-phase supplies to domestic installations.

Section 6 – Electrical supply considerations

Table 6.3   Example diversified load data for a three-storey townhouse

Way	Module description	Maximum demand per circuit (kW)	07.00–13.00		13.00–19.00		19.00–01.00		01.00–07.00	
			Operational hours	Electrical load (kW)	Operational hours	Electrical load (kW)	Operational hours	Electrical load (kW)	Operational hours	Electrical load (kW)
A	Surge protection									
B	PV isolator (interlock with mains isolator)									
C	Main isolator									
D	EV charge point	7.0	0	0.0	0	0.0	0	0.0	6	7.0
E	ASHP	4.0	0	0.0	4	4.0	4	4.0	0	0.0
F	Hydrobox	2.0	0	0.0	4	2.0	4	2.0	0	0.0
G	Backup heater (underfloor heating)	3.0								
H	Booster heater (domestic hot water)	3.0								
I	Kitchen hob	3.0	0	0.0	1	3.0	0	0.0	0	0.0
J	Kitchen oven	6.0	0	0.0	1	6.0	0	0.0	0	0.0
K	Lighting: ground floor, garage, exterior	0.4	0	0.0	1	0.4	1	0.4	0	0.0
L	Lighting: first and second floor	0.4	0	0.0	1	0.4	1	0.4	0	0.0
M	Sockets (ring final circuit): garage	3.0	1	3.0	0	0.0	0	0.0	0	0.0
N	Sockets (ring final circuit): ground floor	1.0	3	1.0	3	1.0	3	1.0	0	0.0
O	Sockets (ring final circuit): first and second floor	1.0	0	0.0	0	0.0	4	1.0	0	0.0

Table 6.3 cont.

Way	Module description	Period of use – Maximum demand per circuit (kW)	07.00–13.00 Operational hours	07.00–13.00 Electrical load (kW)	13.00–19.00 Operational hours	13.00–19.00 Electrical load (kW)	19.00–01.00 Operational hours	19.00–01.00 Electrical load (kW)	01.00–07.00 Operational hours	01.00–07.00 Electrical load (kW)
P	Sockets (ring final circuit): kitchen	2.5	1	2.5	1	2.5	2	2.5	0	0.0
Q	Mechanical ventilation with heat recovery	1.0	3	1.0	3	1.0	3	1.0	0	0.0
R	Smoke and heat detectors	0.1	6	0.1	6	0.1	6	0.1	6	0.1
	Totals (kW)	31.40		7.60		20.40		12.40		7.10
	1Φ supply requirement (amps)			33.0		88.7		53.9		30.9
	Balanced 3Φ supply requirement (amps)			11.0		29.4		17.9		10.2

Section 6 – Electrical supply considerations

It is important that the engineering design of the individual installation considers the maximum demand required. A larger housing development with several properties could use the lower figure to assess the aggregated demand across the whole site and to be connected to the same 11 kV transformer.

For the electrified heat estimates only the active circuits are considered for the diversified load. The backup and booster heating are ignored because they will only be in use when the main systems are off.

It should be noted that the example is for a three storey new build townhouse that had deployed a heat pump and underfloor heating. The efficiencies of the heat pump allow less electricity to be used. In a house with electrical heating using more traditional resistance elements (for example, storage heaters that are not on a separate overnight tariff), the peak electricity demands could be significantly higher and a three-phase supply may be required also.

6.1.5   Circuit protection

Circuits for electrified heat should be fed by way of individual radial circuits from the nominated consumer unit or distribution board. Radial circuits will ensure adequate circuit protection specifically for the heating system. It will also improve the resilience of the electrified heating systems because dedicated radial circuits are less likely to be affected by faults in other parts of the electrical installation.

It should be recognized that dual radial circuits are likely to be needed for each part of any electrified heating circuit, one to provide the primary heat pump supply and another to provide for the secondary booster (or backup) system.

Deriving power supplies from a spur connection from existing ring final socket circuits is not recommended as there is the potential to overload those circuits at times of peak demand. For retrofit projects, any existing consumer units should be replaced to ensure there are sufficient ways to supply the radial circuits for electrified heating systems and the associated backup systems.

Before replacing any consumer unit or distribution board, a professional electrical technician or engineer should review the installation and the supply to check that installing a replacement consumer unit is feasible. This may also involve upgrading from a single-phase supply to a domestic three-phase supply.

For all electrical loads, it is essential that the installation complies with the safety aspects of BS 7671. Correctly calculating cable sizes to match the rated electrical load and to overcome voltage drops will ensure compliance with those safety aspects and help electrical efficiency.

Undersized cables run the risk of overheating under operating conditions. This increased temperature is wasted energy, and under extreme conditions the safety of the electrical installation can be compromised.

DNOs may ask for full operating characteristics of any electrified heating system. This may include the assessment of any power factor, and any kW or KVA data, of the electrical equipment.

Protective devices shall be selected in accordance with the requirements of BS 7671. Protective devices should be compatible with the consumer unit or distribution board, in accordance with the consumer unit or distribution board manufacturers' instructions. This process should include consideration of:

(a) the current rating of devices being checked against the maximum demand of the heating system;
(b) the protective device characteristics (type B or type C devices) should match the start up requirements of the electrified heating system and any integral pumps or similar equipment;
(c) the use of MCBs or RCBOs;
(d) the recommended use of arc fault detection device (AFDD) in addition to (and often combined with) MCB or RCBO; and

Section 6 – Electrical supply considerations

(e) the installation of any surge protection to protect the electrified heating system and associated controls from adverse electrical supply conditions.

The location of any isolation devices for both internal thermal storage units and external heat pump units should also be assessed. For external isolation devices, the index of ingress protection rating should be checked.

6.1.6 Power quality

Power quality is an essential component in the energy efficiency of an electrical installation.

Electronic control systems for electrified heating systems may include specialist pump controls. These controls usually allow the equipment to operate more efficiently but can also cause harmonics or flicker. There is an obligation for end users to ensure that the equipment connected to a supply does not have a detrimental effect on the wider grid. Limits are imposed on acceptable levels of harmonics and flicker on any electrical installation.

Applications for new supplies request the applicant to declare that their electrical equipment complies with the standards shown in Table 6.4, to reduce the risks of harmonics and flicker.

Table 6.4   Harmonics and flicker standards

BS EN standard	IEC standard	Title
BS EN 61000-3-2	IEC 61000-3-2	Limits for harmonic current emissions (equipment input current $\leq$ 16 A per phase).
BS EN 61000-3-3	IEC 61000-3-3	Limitation of voltage changes, voltage fluctuations and flicker in public low voltage supply systems, for equipment with rated current $\leq$ 16 A per phase and not subject to conditional connection.
BS EN 61000-3-11	IEC 61000-3-11	Limitation of voltage changes, voltage fluctuations and flicker in public low voltage supply systems, for equipment with rated current $\leq$ 75 A and subject to conditional connection.
BS EN 61000-3-12	IEC 61000-3-12	Limits for harmonic currents produced by equipment connected to public low voltage systems with input current >16 A and $\leq$ 75 A per phase.

6.2   Engaging with DNOs and suppliers

6.2.1   The National Grid and DNOs

The National Grid is an interconnected network that delivers electricity from suppliers to consumers. It ensures that all consumers have a supply of electricity to meet UK demand.

At a local level, the DNOs provide:

(a) single-phase or three-phase electrical connections for new developments; or
(b) electrical supply upgrades for refurbishments (larger current ratings or swap out from single-phase to three-phase).

Section 6 – Electrical supply considerations

The extent of the works provided can either be simple and low cost (for instance a short low-voltage cable that needs upgrading on a single site) or complex and expensive requiring new high-voltage distribution cables, transformer substations and low-voltage supply cables to the final point of connection.

There are 14 licensed DNOs in the UK and each is responsible for a regional distribution services area. Ofgem regulate the distribution and transmission networks. The Energy Networks Association (ENA) is an industry body formed by the DNOs of both electricity and gas and provides a full list of DNOs on its website (https://www.energynetworks.org/).

6.2.2 Applications for upgrades and new supplies

The ENA website contains a publications section with details of processes and additional information concerning the connection and integration of low-carbon technologies including electrified heat, and specifically heat pumps. The ENA website also includes a chart titled, the *Electric Vehicle Charge Point and Heat Pump Connections Process*. This chart describes the process followed for varying sizes of heat pumps and details the approach if a new installation requires:

(a) just heat pump loads;
(b) both heat pump and EV charge point loads; and
(c) just EV charge point loads.

When applying for new supplies, the DNOs require that a declaration is made of any EV charge points, heat pumps, energy storage or local electrical generation. A special section detailing these low-carbon technologies is often included in the application form. The following application form (see Figure 6.2) is an example derived from UK Power Networks (UKPN) and similar forms are available on most DNO websites.

Figure 6.2 Low-carbon technologies data input (from UKPN)

Section H: Tell us about any low carbon technologies

Such as generation, storage, heat pumps or EVCP being installed as part of your development

Please indicate which low carbon technologies you are installing as part of your development (tick all that apply)

☐ Generation   ☐ Storage   ☐ Heat Pumps   ☐ Electric vehicle charging points   ☐ Other

Please also confirm relevant ENA generation application forms (technical sections only) as below, and submit with this form.

– Individual installations <3.68 kW use G98 Form B
– Multiple installations <3.68 kW use G98 Form A
– Individual and multiple installations >3.68 kW <50kW use Form A1-1
– Individual and multiple installations >50 kW use standard application form (SAF) part 3 only. Parts 4 and 5 may subsequently be required

Further details on the specific heat pumps will be required (see Figure 6.3). It is recommended that the manufacturer is consulted also as a database of heat pumps on the market is kept by the ENA. The DNO will require that the load characteristics of the heat pump are declared as part of the new or upgraded supply application.

In addition to the model manufacturer and model name, the installer should declare whether the heat pump:

(a) is used for heating only or heating and cooling; and
(b) has any form of backup or booster heating.

Figure 6.3 Heat pump characteristics electrical supply application form (Derived from UKPN)

H2. Heat Pumps

You may complete this section or alternatively append a completed ENA form with this application.

Power Quality Declaration – Heat Pumps Only

Heat Pump Manufacturer

Heat Pump Model

How will the Heat Pump be used? (please tick one of the options)

- ◯ The heat pump model state will provide HEATING ONLY
- ◯ The heat pump model state will provide HEATING & COOLING

Does the Heat Pump have additonal components installed?

◯ Back-up heater – on-board ◯ Back-up heater – external ◯ Booster heater – on-board
◯ Booster heater – external ◯ Immersion heater – on-board ◯ Immersion heater – external

Is this model in the ENA Heat Pump Type Register Database and is the information in the Database correct?
See register in database found in the second paragraph under 'Processes & Forms' on the ENA website here. If yes please proceed to 'Declaration'.

◯ Yes – register No. ◯ No

If 'No' please fill in the following additional Power Quality details required for non-registered Heat Pump Models.

Datasheet and other Power Quality documentation for the Heat Pump attached to this application? ◯ Yes ◯ No
Must be provided. It is the installer's responsibility to ensure all information required to populate the Heat Pump Type Register Database is provided.

Does the installation meet the Microgeneration Certificate scheme Product requirements? ◯ Yes ◯ No

Harmonics
Does the proposed installation comply with the technical requirements of BS EN/IEC 61000-3-2? ◯ Yes ◯ No

Harmonics
Does the proposed installation comply with BS EN/IEC 61000-3-12? ◯ Yes ◯ No

Flicker
Does the proposed installation comply with the technical requirements of BS EN/IEC 61000-3-3? ◯ Yes ◯ No

Flicker
Does the proposed installation comply with BS EN/IEC 61000-3-11? ◯ Yes ◯ No

Please refer to the Manufacturers Declaration of Conformity, device type text certificate and datasheet. If using the multiple installations spreadsheet, the confirmation of standards compliance should refer tot he whole installation, i.e. at the point of common coupling.

* https://microgenerationcertification.org/mcs-standards/product-standards/heat-pumps/

Declarations should also be made on the power quality aspects of the heat pump, particularly relating to harmonics and to flicker.

Following submission of the application, an assessment will be made by the DNO to clarify the impact that low-carbon technologies at one installation may have on the wider network. It may mean that specific operating conditions are imposed on the new connection.

Each electrical installation has a unique identifying number known as a meter point administration number (MPAN), or meter point reference number (MPRN) in Northern Ireland. It is allocated when new applications are made on new sites. For existing sites, it can be found on the customer's bill or by contacting the energy supplier and will be needed as part of any supply upgrade. Figure 6.4 details the numerical information that makes up the MPAN.

Figure 6.4 Example MPAN information

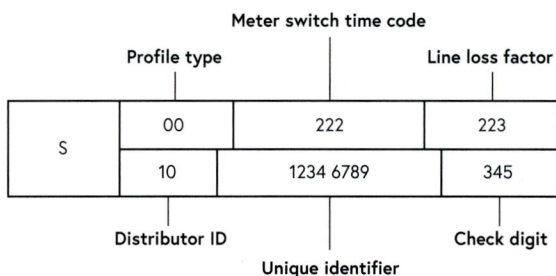

6.3 Electrified heat and smart grids

Electrical supplies to domestic installations often use one meter. Billing to a property has typically consisted of a standard tariff that changes infrequently. Some properties may have used a second supply to provide off-peak energy for electrical storage heaters on an Economy 7 tariff (see Table 6.5).

Table 6.5 Domestic meters and tariff types

Domestic meter type	Domestic tariff	Commentary
Standard analogue meter or standard digital meter equivalent.	Standard rate	Single meter and constant tariff over a 24-hour period (common consumer unit). No incentive to save energy or track usage.
Standard analogue meters or standard digital meter equivalents.	Standard rates (one constant, one off-peak)	Dual meters (with separated consumer units and circuits). Different tariffs for each electrical service over a 24-hour period. Normal consumer unit has constant rate. Economy 7 consumer unit has cheaper overnight rate for electric heating and hot water. No incentive to save energy or track usage.
Smart meter (Smart Metering Equipment Technical Specifications (SMETS) 1).	Standard rate	Single meter and constant tariff over a 24-hour period (common consumer unit). Encourages energy saving through behaviour change.

Smart grid concepts encompass grid generation and distribution, local smart meters (see Figure 6.5), controls and load management at the point-of-use. It allows for a two-way flow of both energy and information on the consumption and generation of electricity.

At the heart of the smart grid is an intelligent control centre that controls generation and distribution. The control centre carefully controls energy generation and prioritizes which generators operate. Smart interaction occurs when individual equipment is powered up or down according to the status of the smart grid.

For domestic users, a smart grid can prioritize loads such as vehicle to grid (V2G) EV charge point or restrict the demand of other loads such as heat pumps. The IET publication *Code of Practice for Electric*

Section 6 – Electrical supply considerations

Vehicle Charging Equipment Installation[11] provides more information on V2G, smart infrastructure integration and installations in dwellings.

Figure 6.5   A domestic smart meter

Smart energy meters can also allow the use of flexible tariffs to be implemented (see Table 6.6). Agile tariffs for electrical energy consumption have entered the domestic housing energy market. This system alters the cost of electrical energy depending on when it is used. Costs vary continuously throughout the day and are significantly higher at designated peak times than off-peak times.

Smart meters feature a display screen that shows how much electricity is being used and its cost in real time. Readings are also automatically sent to energy suppliers at least once a month to avoid estimated bills.

The use of energy smart meters can encourage better energy saving behaviour because users can see how much energy is being used at a particular time and manually switch off unnecessary appliances and loads.

Economy 7 and similar off-peak tariffs encouraged the use of storage heaters and water cylinders to make use of cheap-rate electricity. Modern versions supply space and water heating through off-peak tariffs, time-of-use tariffs, demand-side response and flexibility services, contributing to network flexibility in similar ways to EVs and batteries.

Modern forms of energy storage include large battery units to store electricity at cheaper tariffs for use on the premises during peak times. V2G is another form of battery energy storage and uses two-way energy flows from EVs. This discharges electrical energy into the building whilst the vehicle is parked and then recharges again overnight.

Electrified heating solutions can be designed and operated to incorporate thermal and/or electrical storage so that overnight cheap-rate charging can be made for daytime use in traditionally expensive periods.

6.4   Smart integrated design

6.4.1   Integration concepts and principles

Electrified heat is just one of several loads to be found on modern domestic consumer units. If all the electrical loads were on simultaneously, it would overload the intake. Load diversity does mitigate this

---

[11]IET *Code of Practice for Electric Vehicle Charging Equipment Installation, 4th Edition* IET, London, 2020.

risk, but any load controls are limited to that load. There is little co-ordination and it is down to theoretical chance that not all loads operate simultaneously. Integration of loads across the whole installation provides a greater level of certainty that the maximum capability of the supply will not be exceeded, whilst still allowing all loads to operate satisfactorily.

Figure 6.6 shows electrified heat as a typical heat pump system. The heat pump system has a main heat pump, a backup heater and thermal storage for the DHW and for the space heating (SH). There is also an EV charge point, PV microgeneration and on-site electrical energy battery storage.

Table 6.6 Domestic smart meters

Domestic meter type	Domestic tariff	Description
Smart meter (SMETS2 or Secure TM SMETS1).	Half-hourly reading and flexible rate.	Single meter with varying electricity prices updated daily and based on wholesale costs. Encourages load shifting away from 16.00–19.00 peak hours. Equipment should have time controls to switch off during potentially expensive periods. (common consumer unit)
Smart meter (SMETS2 or Secure TM SMETS1).	Half-hourly reading and two set rates.	Single meter offering limited period of low-cost off-peak electricity overnight. Normal tariff outside this period. Useful for overnight charging EVs or electrified heating. Equipment will need time controls to switch on during cheaper periods. (common consumer unit)

Figure 6.6 Electrical requirements for integrated design of a heat pump system

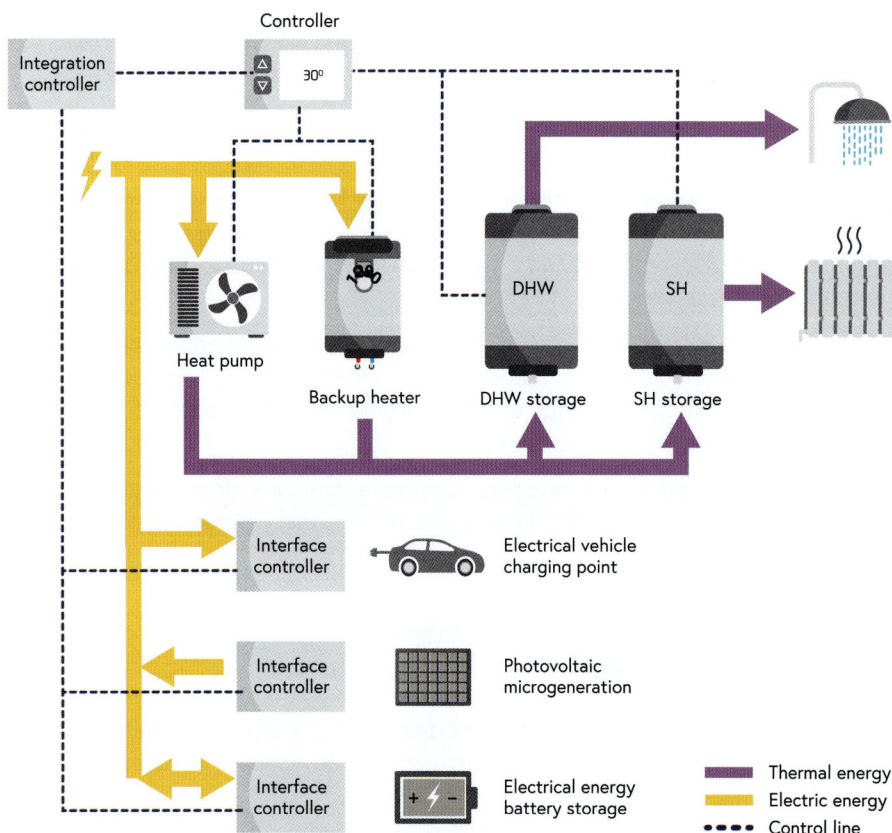

Section 6 – Electrical supply considerations

Each system will have their own controls. Designing interfaces and links between the systems allows the loads to be coordinated so they run seamlessly with no risk to the intake through overload of the installation.

6.4.2 Electrified heating thermal storage

Electrified heating systems should, where possible, include a form of thermal storage. There are many benefits for thermal storage that include:

(a) providing a stored heat backup should the system fail that provides additional hours of heat;
(b) enabling the sizing of the heat generating plant, such as heat pumps, to be optimized and reduced;
(c) enabling load matching between heat source and heat emitters;
(d) enabling demand side response capability; and
(e) enabling the provision for backup electric immersion heaters to be installed.

This approach will add resilience to the heating system should power failures occur.

This will enable the building to participate with smart grids and operate within demand-side response initiatives.

6.4.3 Building electrical energy battery storage

This technology is still gaining a foothold in the domestic marketplace. Configured correctly, battery storage will allow energy from PV microgeneration and other technologies to be stored locally and then used at peak times. With the correct tariff and smart meter connections, battery storage can be used to store off-peak, cheaper electricity for use at later, more expensive, peak times to operate electrified heat and other larger loads.

Lithium-ion's current 10-year warranty/cycle life is available with fully integrated low risk solutions. Lithium batteries (see Figure 6.7) are generally safe and unlikely to fail, but only if there are no defects

Figure 6.7 Lithium-ion battery construction

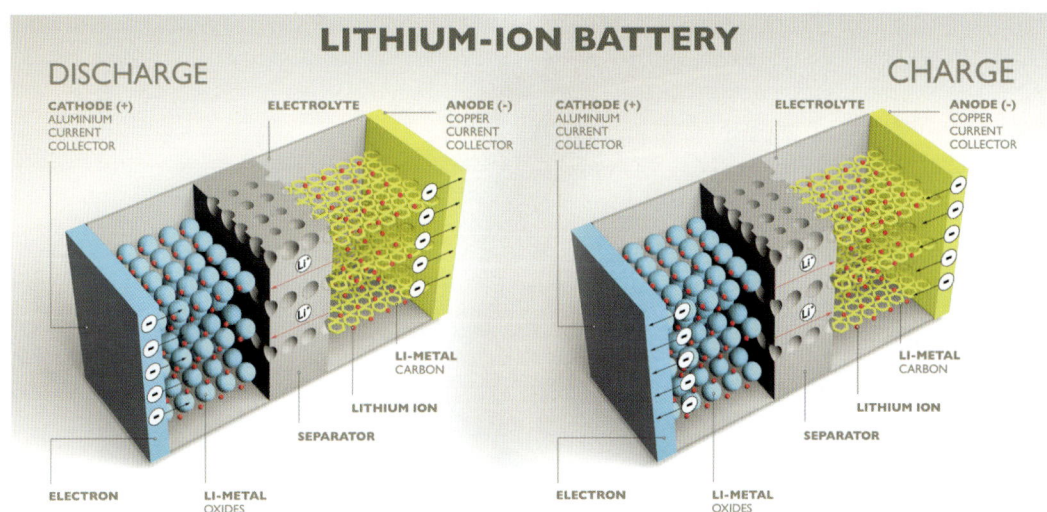

and the batteries are not damaged. However, when lithium batteries fail to operate safely (or are damaged), they may present a fire and/or explosion hazard.

The IET publication *Energy Storage for Power Systems* considers risk management in greater detail[12].

6.4.4 PV microgeneration

PV is often added to electrical installations and can supply locally generated electricity during daylight hours to provide electrical power either to the local installation or to be exported through the intake to the grid. A roof-mounted PV array is shown in Figure 6.8.

PV is worth considering if there is sufficient roof space and the structural integrity of the roof is able to support an installation.

If correctly configured, electrified heat can be supported in part by PV. During times of low demand, the PV installation can charge batteries for discharge later. When not required, the generated electricity can be exported onto the national grid attracting payment.

Figure 6.8 A roof-mounted PV array

[12]Ter-Gazarian, AG *Energy Storage for Power Systems, 3rd Edition* IET, London, 2020

Section 6 – Electrical supply considerations

PV panel outputs vary from around 250 W to over 400 W per panel. PV panel sizes vary from around 1.6 x 1 m to 2 x 1 m per panel and weigh around 18–22 kg each. At present PV panel efficiencies can vary between 18–24 % and it is expected that PV panel efficiencies will improve by 2025 to achieve 28–30 % efficiency.

A typical array of 250 W PV panels, measuring 2 m^2, could provide:

(a) 1 kWp using four panels measuring 8 m^2 with an annual output of 850 kWh per year; or
(b) 4 kWp using 16 panels measuring 28 m^2 with an annual output of 3,400 kWh per year.

The IET *Code of Practice for Grid-connected Solar Photovoltaic Systems* considers these installations in greater detail[13].

6.4.5 EV charge point

An electrical installation with electrified heating systems may also include an EV charge point. Co-ordination is recommended between the relatively large and sustained loads for electrified heat and EV charge points to avoid overloads when they are both connected simultaneously. This will involve assessing the priority loads and restricting other loads for peak periods. Typically, an installation may well be heat-led as a priority and EV charging may have to wait until heat is stored for use.

Some installations may also use V2G or vehicle to home (V2H) connections. This uses the EV battery as an additional energy storage device to support the electrical load of the building. The benefits of this include:

(a) when plugged in during the day and night, EVs provide two-way charging and discharging with the heating system and act as supplementary building energy storage; and
(b) the maximum demand of electrified heat will benefit from capacity benefits of batteries, including EVs.

The UK government's Climate Change Act has set a goal of being net zero carbon by 2050. This plan includes a ban on selling new petrol and diesel cars in the UK from 2030. The inclusion of EV charging can futureproof installations and encourage EV uptake, which in turn reduces air pollution and the risks associated with climate change.

EV charge times will vary according to the EV battery type, the battery capacity, the existing level of charge and the charger used. There are two different types of EV charging for domestic use:

1. slow: charging station typically rated at 3 kW and providing an 8–10-hour charge, which is ideal for charging at home overnight; and
2. fast: charging station typically rated at 7 kW providing a 4-hour charge.

Figure 6.9 shows a fast EV charging station.

The IET *Code of Practice for Electric Vehicle Charging Equipment Installation* considers these installations in greater detail.

---

[13]IET *Code of Practice for Grid-connected Solar Photovoltaic Systems* IET, London, 2015

Figure 6.9   A wall-mounted fast EV charging station

6.5   Summary

(a) Always check the supply capacity for a property to ensure it is suitable for additional electrified heat and other new loads.

(b) Ensure that suitable diversity factors are applied when determining overall supply capacity for all new and existing electrical loads.

(c) When needed, electrical supply upgrades can be easily arranged with the DNO.

(d) Consider futureproofing the property with PV, EV charging and battery storage spare ways in consumer units.

(e) Electrical solutions should be designed and installed to the requirements of BS 7671.

(f) Electrical works should be carried out by a qualified electrician to Part P of the Building Regulations.

▬ Section 7

The business case for electrified heat

This Section will discuss the business case for electrified heating systems available to the domestic market today.

Consideration is given to the drivers, benefits, risk considerations along with the methodology in gathering information to financially assess the merits of electrified heating. Finally, a worked example is presented for a fictitious case.

7.1 Introduction

7.1.1 Drivers

Unfortunately, the decision to replace an existing boiler usually occurs when it fails. It is likely that this will happen when it is most needed – usually during a cold spell – when there is an urgency to get the boiler replaced as quickly as possible. This puts the householder under pressure to replace the boiler with a like-for-like replacement. The decision to upgrade an existing heating system to one with a heat pump is more of a challenge and a longer lead time is needed to consider the implications.

The driver for electric heating in new buildings is more straightforward. In the *Future Homes Standard* (currently under consultation), the UK government plans to make changes to Part L and Part F of the Building Regulations to improve the energy efficiency of new homes. The new *Future Homes Standard* should ensure that all new homes built from 2025 will produce 75–80 % less carbon emissions than homes under current regulations. Homes built in 2025 will all be heated using non-fossil fuels making electrified heating an obvious and cost-effective choice.

The UK government is also promoting the uptake of heat pumps as a contributing solution towards meeting their decarbonization goals, with ambitions for the industry to be installing 600,000 heat pumps a year by 2028. The CCC has already set a target of one million heat pumps a year by 2030 towards an eventual total of 19 million to achieve the government's net zero targets. There are currently around 16,000 heat pump installations in the UK every year, but a new report from the CCC has proposed that this figure rises to almost 50,000 between 2021 and the end of the decade, rising to one million a year thereafter.

7.1.2 The advantages and disadvantages of electrified heat

One of the main advantages of electrified heat is decarbonization, but it also has the potential to improve thermal standards, operation and control (leading to reduced energy consumption and carbon emission savings). Another advantage of electrified heat is a reduction in household air pollution from combustion flue gases, that include carbon monoxide (CO), nitrogen dioxide (NO_2) and nitrogen oxides (NOx). These gases are removed at source from the dwelling and will immediately improve the air quality surrounding the home.

The main disadvantage with electrified heat is the cost of energy. If a fuel switch is made from natural gas to electricity, the unit charges will be higher (mainly due to renewable energy charges on electricity bills) and careful analysis of operating costs is required to understand the true economic picture.

The UK government has a number of subsidies and grants to support the purchase and operating of heat pumps, but these are prone to rapid change, rebranding and can sometimes be scrapped completely.

Section 7 – The business case for electrified heat

Historic schemes, such as the Green Deal, Feed-in Tariffs and Green Homes Grant have left many participants disillusioned. There is no one scheme that will work for all and all schemes should be carefully assessed for each individual case.

Manufacturers and installers will benefit hugely from the market leap towards heat pump technology. The renewables market size between 2015–2020 was approximately £19 billion and it is estimated that this will increase to £110 billion between 2025–2030. In line with the Sustainable Development Scenario, the share of clean heating technologies, including heat pumps, district heating, renewable and hydrogen-based heating, needs to more than double to 50 % of sales by 2030 (sales are currently only 10 %).

The main advantages of electric heat are:

(a) CO_2 reduction;
(b) reduction in local air pollution;
(c) supporting the market leap; and
(d) benefiting from subsidies and grants

The main disadvantages are:

(a) the high cost of electricity relative to gas;
(b) the potential for changing policies concerning heat pumps; and
(c) higher capital expenditure (for heating unit, long thermal length radiatiors and heat demand reduction measures).

7.1.3 Stakeholders

There are several important stakeholders to consider when developing a business case for electrified heat. These include the customer, installers, manufacturers, the government (particularly BEIS), utility companies and local authorities. No one stakeholder remains static in the business case for electrified heat and they should all be carefully considered.

The most important stakeholder is the customer and they need to be aware that it is more expensive to install, and potentially operate, a heat pump as opposed to a gas boiler. It is also important to explain the benefits of installing a heat pump to the customer, and make them aware that most of the investment relies on incentives and grants offered by the government and local authorities. The customer should understand that the heat pump operates differently to a gas boiler and the impact that this has on heating temperatures and hot water provision. It is also worth checking on local authority planning conditions, which could either aid or conflict a customer's electrified heat projects (such as joining district heating schemes).

A good and up-to-date understanding of the marketplace is needed as manufacturers innovate regularly and price changes can occur rapidly. Finally, monitoring electricity supply companies is important as rates of energy change and can reduce if the solution includes off-peak electricity tariffs. Additionally, from time to time the National Grid pilots a project to utilize excess generated electricity from local renewables and this can also benefit the electrified heating business case.

7.1.4 Key design considerations and risk management

PAS 2035/2030 *Retrofitting dwellings for improved energy efficiency. Specification and guidance* has been developed to address and counter risks associated with extensive retrofit projects. It is likely that any publicly funded installations will need to be installed within this scheme. This would also require the installer to be registered under PAS 2030.

Section 7 – The business case for electrified heat

The principle technical considerations include an assessment of the building, especially if the building is a retrofit project. These are shown in Table 7.1.

Table 7.1 Building considerations in retrofit projects

Aspect	Technical consideration	Outputs/options
Building fabric U-values	Determination of thermal performance of the building. Does it require improvement?	Double- or triple-glazing. Roof insulation. Cavity wall insulation. Underfloor insulation. New external doors.
Building fabric air leakage	Determination of envelope air leakage. Does it require improvement?	Repair of fabric holes. Sealing at fabric junctions. Double- or triple-glazing. New external doors.
Building electrical supply	Determination of the electrical supply capacity. Can it support the additional connection?	Power supply upgrade. Final circuit to feed the heat pump. Load control.
Location for heat pump	Location of the heat pump external unit, internal unit, thermal store and hot water system. Is there enough space?	Space required displacing existing cupboards and/or garden.

Similarly, an assessment of the incumbent or proposed heating system itself is required. Again, this is particularly important if the building is a retrofit project. Considerations are shown in Table 7.2.

Table 7.2 Heating system considerations in retrofit projects

Aspect	Technical consideration	Outputs/options
Heat source type	Appreciation of the type and application. Does it provide heating and hot water? How noisy is the system?	Strategy for heating and hot water. Locating unit to avoid noise issues.
Operating temperatures	Appreciation of the system operating temperature. If operating temperature is 80 °C then what is the impact of a reduction to 55 °C?	Strategy to address a temperature reduction. Emitter resizing. Fabric improvements.
Heating pipework	Appreciation of the pipe sizes, limitations (if existing pipes) and long legs preventing even heat distribution. How well is the existing pipework insulated in roof spaces and voids?	Pipework alterations. Inclusion of commissioning and/or balancing valves. Thermal insulation in voids.
Heat emitters	Appreciation of how the heat emitters were originally sized. Were they originally oversized? Has the building been improved since to reduce heating demands? Will existing emitters heat rooms adequately at a reduced temperature of 55–60 °C?	Determination of current heat losses in all rooms compared against emitter outputs at reduced temperatures. Potential replacement of emitters with larger surface areas to meet heat losses.

All technical risks should be clearly identified and carefully considered. Customers should understand the risks they are taking. For example, if additional insulation and double glazing are not installed, the heat pump may not provide enough heat to warm the house.

7.2 Business case information required

The evaluation for providing electrified heat will require information about the building and its energy performance to establish a baseline case for comparison. This will also require utility information, cost information and an understanding of the available grants and subsidies.

The majority of installations in the UK are concerned with gas boilers being replaced with electric heat pumps. The methodology is the same if the baseline or existing heating system is fuelled from heating oil, liquified petroleum gas or any other types.

7.2.1 Baseline energy consumption and cost

Baseline energy and consumption costs are obtained in existing buildings by obtaining energy data for a 12-month period (from a customer's billing and consumption data). For new houses, this information can be obtained from the building design thermal model (the SAP model). This data will provide monthly and annual gas consumption in kWh showing how heating, hot water and cooking use energy. It is important to estimate how much energy is used by each end use. Hourly information is ideal as this can identify peak demands and loads and is useful in optimum sizing.

Baseline energy unit cost information is also needed for cost modelling and will require electricity and oil, liquified petroleum gas and gas tariff unit rates. These will be available from the customers' bills or in the case of in new builds, by a market survey. When assessing utility prices note that electricity and gas costs are likely to rise and it is recommended that, for cost modelling purposes, rates are increased to account for these future price rises. Flexible time-of-use tariffs, however, may offer a viable pricing solution if correctly configured.

Typical energy costs for 2021 are:

(a) gas at £0.03/kWh
(b) electricity at £0.16/kWh (with a £0.19/kWh price cap)
(c) electricity Economy 7 rate at £0.08/kWh (between 11 p.m. and 6 a.m.)

Energy consumption and cost key points:

(a) energy and consumption costs are highly variable and should be obtained on a case-by-case basis; and
(b) the Economy 7 tariff allows customers to use cheaper tariffs during the night and this could have an impact on the business case if thermal storage is built into the solution.

7.2.2 Annual servicing costs

Both boilers and heat pumps require annual servicing. Actual servicing costs can be obtained from the homeowner if the project is a retrofit. The proposed manufacturer will also be able to provide an accurate estimate for annual servicing costs.

Typical annual servicing costs in 2021 are:

(a) gas boiler: £75 per year; and
(b) electric heat pump: £100 per year.

Warranties and guarantees that manufacturers offer should be included in the business case, to assess the product life cycle. Ideally, a 10-year warranty should be provided and an expected life for both a gas boiler and a heat pump being around 15 years. Each manufacturer, however, will have different warranties.

Section 7 – The business case for electrified heat

7.2.3 Carbon factors

Carbon factors change each year as the grid is further decarbonized (see the UK government's publications for GHG reporting conversion factors https://www.gov.uk/government/publications/greenhouse-gas-reporting-conversion-factors-2020).

Typical carbon factors for use in 2021 (based on 2020 emissions) are:

(a) gas at 0.20374 kg CO_2/kWh; and
(b) electricity at 0.25319 kg CO_2/kWh.

However, the SAP for the energy rating of dwellings methodology uses different carbon factors. The current version is SAP 10.1 that uses:

(a) gas at 0.210 kg CO_2/kWh; and
(b) electricity at 0.136 kg CO_2/kWh.

The SAP electricity emission carbon factor is much lower as it is based on predicted average electricity factors between 2020 and 2024, which favours heat pumps.

7.2.4 Capital cost

The capital cost for an installation is highly variable since it is dependent on the project and the cost of any improvements in terms of thermal insulation and power supply upgrade. It is recognized that these costs could reduce as heat pumps become more widespread and volumes increase.

The capital cost should be fully inclusive and include all design considerations, such as the heat pump installation, power supplies, ancillary works, thermal improvements, controls, commissioning, customer training and balancing.

Cost build-ups should include all materials, labour, plant, access, overheads and profit and VAT. Costs should be obtained from manufacturers, suppliers and wholesalers and installers should use the approved industry rates. Ideally, the capital cost should include an itemized cost build-up so that the customer understands what is included.

Typical basic guideline costs (including VAT) for a three-bedroom house central heating system in 2021 are:

(a) new build:
  i. gas combi boiler and radiator complete installation = £5,000;
  ii. ASHP and emitters complete installation = £8,000; and
  iii. GSHP and emitters complete installation = £20,000;
(b) refit:
  i. gas combi boiler = £2,500; and
  ii. ASHP = £8–10,000.

7.2.5 Incentives and grants

Incentives and grants are available to support the installation of decarbonized heat and include ASHPs and GSHPs. Subsidies and grants change quickly so it is important to understand the incentives and grants on offer at the time of developing a business case – and when any benefits are likely to be phased out.

Renewable Heat Incentive scheme

The RHI is the government's financial incentive for heat pump owners and is run by Ofgem. It was launched in April 2014 and is scheduled to end on the 31 March 2022. Owners who join the scheme and comply to its rules receive quarterly payments for seven years, with the amount dependent on the estimated clean, green renewable energy their system produces.

To qualify for the RHI scheme, the benefactor must be the owner of the installed heat pump, and the heat pump must be an eligible renewable heating solution, specifically a GSHP or ASHP.

In 2021, domestic RHI scheme payments are based on heat pump heat output and are:

(a) ASHP £0.1085/kWh; and
(b) GSHP £0.2116/kWh.

Clean Heat Grant

The Clean Heat Grant is the proposed replacement for the RHI Scheme and is scheduled to commence in April 2022. The Clean Heat Grant is a government proposed scheme to provide homeowners with funding towards installing low-carbon heating systems, such as heat pumps. It is envisaged that homeowners who successfully apply could receive a flat rate payment of £4,000 via a voucher scheme.

The scheme would require eligible households to have high performance insulation already in place along with a valid EPC.

7.3 Business case evaluation methodologies

7.3.1 Simple payback

Simple payback refers to the amount of time it takes to recover the cost of an investment or how long it takes for an investor to reach breakeven.

Simple payback is defined as total costs divided by annual savings, which result in a payback period measured in years.

The simple payback for a heat pump is calculated as:

$$Pb = \frac{C}{S}$$

where:

Pb is the simple payback period in years

C is the cost, including all capital expenditure less any grants obtained in year 1 in £

S is the total savings achieved in comparison with a gas baseline, including energy cost savings, annual service savings and benefits from incentive schemes

Section 7 – The business case for electrified heat

It is generally accepted that, when considering heating system improvements, a simple payback of seven years or less is deemed as a worthy investment by the government and public.

The disadvantage of a payback calculation method is that it ignores the time value of money, interest rates, future energy price increases, deflation and it neglects cash flows received after the payback period. However, it is an easy way to compare projects and easy to understand, so may be suitable for use with domestic customers.

7.3.2 Net present value

Net present value (NPV) considers the cost of entering a project including the net cash that flows back in the future. NPV is important because economic forces mean monetary value changes over time. The advantage of the NPV method is that it considers the basic idea that a future UK pound is worth less than a UK pound today.

NPV is calculated as:

$$\text{NPV} = \frac{R_t}{(1+i)^t}$$

where:

    R_t is the net cash inflow-outflows during a single period (t)

    i is the discount rate or return that could be earned in alternative investments

    t is the number of timer periods

In every period, the cash flows are discounted by another period of capital cost. NPV converts future cash inflow into their current value, using the discounted cash flow rate. If the value is negative, then the project should not proceed or more time may be required. However, if it is a positive value, then the project should be considered a viable option. The NPV method also tells us whether an investment will create financial value for the company or the investor, and by how much.

The final advantage is that the NPV method considers the cost of capital and the inherent risk in making projections on the future. In general, a projection of cash flows ten years into the future is less certain than cash flows projected next year. Cash flows that are projected further in the future have less impact on the NPV than more predictable cash flows in earlier periods.

There are easy ways to calculate the NPV using spreadsheets and online tools.

7.3.3 Whole life cycle costing

Whole life cycle costing considers the total cost of the electrified heating system over its lifetime, from concept to disposal, and includes purchase, energy costs, maintenance, operation and disposal. In most cases, the purchase costs are only a small proportion of the cost of operating it.

This type of calculation should be carried out using either bespoke software or Excel spreadsheets and are based on the NPV calculation methodology. The output should compare lifetime costs versus lifetime savings. A negative result should be discounted and a positive result should be considered further.

Whole life cycle costing can be complex. It is perhaps best suited to customers who own multiple properties, such as a housing association, and who will be investing large amounts of money into heat pumps. This approach could also include considerations for the embodied energy and carbon of the heat pump and its associated components.

7.4 Example scenario

This Section considers a fictitious electrified heating scenario relating to developing a business case appraisal for a retrofit heat pump.

Mr and Mrs Reid are an environmentally conscious couple who own a three-bedroom 1970s house. Mr and Mrs Reid would like a straightforward but comprehensive proposal for a heat pump installation for their home.

The house currently uses a 14-year-old gas condensing boiler for heating and hot water (via a cylinder) and consumes 10,000 kWh in gas annually. Mr and Mrs Reid have already double glazed their home, carried out fabric improvements, such as 300 mm of roof insulation, cavity wall insulation and have sealed air leakages. Mr and Mrs Reid already have a solar thermal hot water system and PV installation at their home and have changed all of their lighting to LED. An EV charging point has been installed for charging their electric car and upgraded their electricity supply. They are in the process of replacing all appliances to be AAA rated.

The Reid's investment appetite extends to simple paybacks of up to seven years. It is notable that Mr and Mrs Reid live in a flood risk zone although they have yet to suffer a flood.

At the time of the proposal the RHI scheme is in existence and is offering to pay £0.1085/kWh for the first seven years.

7.4.1 Step one: gather information

The first step is to establish and test key facts about the property, its energy consumption investment attitudes and any other requirements.

Following a series of discussions and a detailed survey it has been established that:

(a) they have the electrical capacity for a heat pump to be installed, requiring only a new consumer unit and final circuits to be wired;

(b) there is a compatibility between the heat pump and the current installed heating solution and the system has been modelled and assessed as oversized, meaning that the heat emitters can operate at a lower flow water temperature of 55 °C;

(c) their annual gas consumption is 10,000 kWh: they cook using electricity so this gas is consumed in connection with heating and hot water;

(d) their gas rate is £0.03/kWh, their annual standing charge is £75 and gas servicing costs £75 each year;

(e) their current electricity rate is £0.16/kWh;

(f) Mr and Mrs Reid have had three quotes to replace their boiler system and the best value quote is £2,250 including VAT; and

(g) Mr and Mrs Reid have had bad experiences with Economy 7 tariffs and are not interested in this, but have expressed that they may reconsider this in the future.

7.4.2   Step two: practicalities

A design solution has been developed that uses a low-noise ASHP (with a seasonal CoP of 3.5) hung from a first-floor gable wall (positioned to avoid any potential flood damage). An internal heat exchanger is provided in place of the existing gas boiler that provides low pressure hot water at between 30–55 °C, for circulation into their existing radiators.

A new hot water cylinder is being provided that integrates the existing solar thermal installation with the ASHP and, as a precautionary measure, a 3 kW immersion heater is included as an emergency back-up. The cylinder is sized to act as a thermal store should Economy / tariffs be of interest in the future. The cylinder is to be installed in the same location as the existing cylinder, making the solution seamless.

The capital cost of the proposed scheme has been estimated at £8,500 including VAT, and a full cost breakdown is available that includes survey and design costs, provision of a new heat pump, hot water cylinder, controls, installation, customer training and commissioning. Annual heat pump service costs are offered at £95 per year.

7.4.3   Step three: economic and environmental analysis

Existing gas heating (baseline scenario)

The annual energy cost for the gas boiler is calculated as:

$$E_c(\text{gas}) = E_n(\text{gas}) \times U_r(\text{gas})$$

where:

$E_c(\text{gas})$ is the annual energy cost of gas in £ to be determined

$E_n(\text{gas})$ is the annual energy consumption of gas at 10,000 kWh

$U_r(\text{gas})$ is the unit rate of gas £0.03/kWh (conversion from pence to pounds by dividing by 100)

$$E_c(\text{gas}) = 10,000\,\text{kWh} \times £0.03 = £300.00$$

It is noted that as the Mr and Mrs Reid do not cook using gas, and that all of this energy is used in heating and hot water, they will no longer require a gas supply if the heat pump option is followed. Therefore, the standing charge of £75 per annum can be added to the baseline scenario.

$$E_c(\text{gas}) = £300.00 + £75.00 = £375.00$$

The baseline annual operating cost is:

$$A_c(\text{gas}) = E_c + S_c$$

where:

$A_c(\text{gas})$ is the annual operating cost to be determined in £

E_c is the annual energy cost at £375

S_c is the annual service cost at £75

$$A_c(\text{gas}) = £375.00 + £75.00 = £450.00$$

The gas boiler baseline annual operating cost is £450.

Proposed heat pump solution

The annual energy cost for a heat pump is calculated as:

$$E_c(\text{elec}) = \frac{E_n(\text{gas})}{SCoP} \times U_r(\text{elec})$$

where:

$E_c(\text{elec})$ is the annual energy cost of gas to be determined in £

$E_n(\text{gas})$ is the annual energy consumption of gas at 10,000 kWh. Note that boiler efficiency is as high as a condensing boiler, so gas consumption remains unaltered for the building heat load.

$SCoP = 3.5$

$U_r(\text{elec})$ is the unit rate of electricity £0.16/kWh

$$E_c(\text{elec}) = \frac{10,000 \text{ kWh}}{3.5} \times £0.16 = £457.15$$

The baseline annual operating cost is:

$$A_c(\text{hp}) = E_c + S_c$$

where:

$A_c(\text{hp})$ is the annual operating cost of the heat pump to be determined in £

E_c is the annual energy cost at £457.15

S_c is the annual service cost at £95.00

$$A_c(\text{hp}) = £457.15 + £95.00 = £552.14$$

The heat pump comparison annual operating cost is £552.14.

Grants and subsidies

There is income to be earned as the RHI scheme applies in this scenario. The RHI payment can be calculated by:

Section 7 – The business case for electrified heat

$$\text{RHI} = E_\text{n}(\text{HPo}) \times U_\text{r}(\text{RHI})$$

where:

RHI is the annual income from the RHI scheme in £ to be determined

$E_\text{n}(\text{HPo})$ is the annual energy output from the heat pump at 10,000 kWh

$U_\text{r}(\text{RHI})$ is the unit rate for the RHI at £0.1085

$$\text{RHI} = 10,000 \text{ kWh} \times \text{£}0.1085 = \text{£}1,085.00$$

The revised annual operating cost is therefore:

$$A_\text{c}(\text{hp} + \text{RHI}) = \text{Heat pump operating cost} - \text{RHI income}$$

where:

$A_\text{c}(\text{hp} + \text{RHI})$ is the annual operating cost of the heat pump to be determined in £

Heat pump operating cost $= \text{£}552.14$

RHI income $= \text{£}1,085.00$

$$A_\text{c}(\text{hp} + \text{RHI}) = \text{£}552.14 - \text{£}1,085.00 = -\text{£}532.86$$

The heat pump baseline annual operating cost including RHI income produces a profit of £532.86.

Overall saving

The overall saving is the difference between the cost of operating the gas boiler and the heat pump:

$$E_\text{s}(\text{tot}) = A_\text{c}(\text{gas}) - A_\text{c}(\text{hp} + \text{RHI})$$

where:

$E_\text{s}(\text{tot})$ is the total energy savings to be determined in £

$A_\text{c}(\text{gas})$ is the annual operating cost of gas at £450

$A_\text{c}(\text{hp} + \text{RHI})$ is the annual operating cost with RHI considered at (£532.86)

$$E_\text{s}(\text{tot}) = \text{£}450 - -\text{£}532.86 = \text{£}982.86$$

The overall operating cost saving is £982.86

Capital cost

Mr and Mrs Reid have been quoted for a gas boiler replacement at £2,250.00 and the determined cost for the new heat pump is £8,500.00. The cost uplift from a gas boiler to a heat pump is therefore:

Section 7 – The business case for electrified heat

$$CUp = HPcost - GBcost$$

where:

    CUp is the cost uplift in £

    HPcost is the heat pump cost at £8,500.00

    GBcost is the gas boiler replacement cost at £2,250.00

$$CUp = £8,500.00 - £2,250.00 = £6,250.00$$

The cost uplift for a heat pump is £6,250.00

Simple payback

The simple payback for a heat pump is calculated as:

$$Pb = \frac{C}{S}$$

where:

    Pb is the simple payback period in years

    C are the costs including all capital expenditure less any grants obtained the first year at £6,250.00

    S is the total savings achieved in comparison with a gas baseline including energy cost savings, annual service savings and benefits from incentive schemes at £982.86

$$Pb = \frac{£6,250.00}{£982.86} = 6.35 \text{ years}$$

The simple payback for the heat pump investment is 6.35 years. It should be noted that the RHI scheme only makes payments for seven years so, after this period, annual costs will increase.

Carbon savings

The formula to determine carbon emissions follows:

$$C_e = A_e \times C_f$$

where:

    C_e is the carbon emissions in kg CO_2 pa

    A_e is the annual energy consumption in kWh pa

    C_f is the fuel carbon conversion factor in kg CO_2/kWh

This is carried out for both the gas and electricity options and then the difference subtracted.

Section 7 – The business case for electrified heat

Gas

$$C_e(\text{gas}) = A_e \times C_f$$

where:

$C_e(\text{gas})$ is the carbon emissions due to gas boiler in kg CO_2 pa

A_e is the annual energy consumption in 10,000 kWh pa

C_f is the gas carbon conversion factor in 0.210 kg CO_2/kWh

$$C_e(\text{gas}) = 10,000 \text{ kWh} \times 0.210 = 2,100 \text{ kg } CO_2$$

Electricity

$$C_e(\text{hp}) = A_e \times C_f$$

where:

$C_e(\text{hp})$ is the carbon emissions for electric heat pump in kg CO_2 pa

A_e is the annual energy input into the heat pump is 10,000 kWh pa/SCoP $= \dfrac{10,000 \text{ kWh}}{3.5}$ $= 2,857$ kWh

C_f is the electricity carbon conversion factor in 0.136 kg CO_2/kWh

$$C_e(\text{hp}) = 2,857 \text{ kWh} \times 0.136 = 388.6 \text{ kg } CO_2$$

The carbon saving is:

$$C_e(\text{save}) = C_e(\text{gas}) - C_e(\text{hp})$$

where:

$C_e(\text{save})$ is the carbon emissions saving in kg CO_2 pa

$C_e(\text{gas})$ is the carbon emissions for gas boiler at 2,100 kg CO_2 pa

$C_e(\text{hp})$ is the carbon emissions for electric heat pump at 388.6 kg CO_2 pa

$$C_e(\text{save}) = 2,100 - 388.6 = 1,711.4 \text{ kg } CO_2$$

The carbon savings for the heat pump option is 1,711.4 kg CO_2 pa.

7.4.4 Step four: commercial arrangements and offer

A full commercial offer can be made to Mr and Mrs Reid. This should contain a summary of the key points and be provided with supporting information including technical details of the heat pump, images and detailed calculations. The submission should include case studies and reference statements (obtained with permission).

The key elements of the commercial offer should include:

(a) **technical details** of the proposed heat pump, installation and operation including layout/location drawings.
(b) **reference data** that has been used (such as carbon conversions, reference cost of replacement gas boiler, energy unit tariff rates).

(c) **environmental benefits.** In this case, carbon savings of 1,711.4 kg CO_2 annual emissions saving, flood risk mitigation as the heat pump is located off of the ground and a reduction in gas boiler emissions including CO, NO_2, and NOx.

(d) **capital cost** along with proposals for the ongoing annual service.

(e) **simple payback**. In this case, 6.34 years and assumes RHI payments (noting that the RHI payments only last for seven years).

(f) **warranty and life** including anticipated life and details of warranties and guarantees, and how long these last.

(g) **risks and opportunities**. In this scenario, the biggest risk is the unit cost of electricity rising, would cause the payback to increase. Therefore, it would be wise to include an estimate on the impact of charging some heat overnight using Economy 7 tariffs. This could be a strategy to reduce a future risk of cost rises. Additionally, each year the carbon emission factor is set to reduce until electricity is carbon-free. This means that carbon emission savings will increase each year. In this scenario, if Mr and Mrs Reid require further persuasion, a NPV calculation could be provided or a whole life cycle cost appraisal that includes both cost and carbon emissions.

7.5 Summary

(a) Gas is a cheap fuel and electricity is an expensive fuel. However, gas is a dirty fuel and electricity is a clean one.

(b) Heat pumps are more expensive to install compared with traditional gas boiler.

(c) Government subsidies make heat pump installations more affordable, but schemes and incentives can change quickly.

(d) Optimal business case conditions can be achieved when buildings are improved (insulate, insulate, insulate and then ventilate).

(e) The lowest heat pump operating costs are achieved when time-of-use tariffs are used, which benefit from half price overnight electricity. This requires a thermal store that is charged overnight and then the heat is released during the day (similarly to a night storage heater).

(f) Cost modelling simple payback methods usually suffice. However, on more complex projects, life cycle modelling can be provided that uses NPV techniques.

Section 8

Future perspectives: Heat as a Service

There are several ways in which households have had their low-carbon heating systems, specifically heat pumps, installed. Most will have bought them outright and will perhaps have received government subsidies. It is anticipated that in the future different schemes and models will emerge for households. This could include 10-year contracts for fixed price per unit of heat used, which are currently operating in countries like Denmark, and include most of the installation and maintenance costs.

In this Section, the future is considered and an optimal Heat as a Service (HaaS) model presented. This concept has been developed by Energy Systems Catapult (ESC) as a method to consider how HaaS could be delivered and how it can help deliver heat pumps (and other low-carbon heating systems) to people's homes.

Energy Systems Catapult (ESC) was set up to accelerate the transformation of the UK's energy system and ensure UK businesses and consumers capture the opportunities of clean growth. It is an independent, not-for-profit centre of excellence that bridges the gap between industry, government, academia and research. The ESC has created a range of service platforms that are a route for innovators and other energy sector stakeholders to work with the ESC and access unique capabilities and assets. The ESC takes a whole-system view of the energy sector, helping to identify and address innovation priorities and market barriers, to decarbonize the energy system at the lowest cost.

8.1 Case study context

Heat pumps present a reliable technology solution to decarbonize domestic heat to help achieve net zero. The number of heat pumps installed in peoples' homes is increasing, but less than 4 % of homes currently have them installed. It is widely accepted that today's consumers have little reason to change their heating system to low-carbon: doing so can be expensive, technically complex and, even if people are interested in transitioning, people are unsure where to receive advice. Those living in social housing may never be given the choice to benefit from the comfort, control and potentially lower cost of low-carbon heating systems, as social housing providers need confidence that heat pumps will keep their tenants comfortable.

The ESC has been researching how people use heat in their homes, and how a new type of energy tariff could increase the uptake of low-carbon heating. This new tariff, called HaaS, has been explored by suppliers and service providers in the past. A few have developed and trialled more advanced HaaS, using smart heating controls that uses data to help the consumer understand how they like to be warm, and inform the provider how to price and deliver HaaS. The ESC has publicly trialled a version of HaaS and enabled some service providers to deliver it themselves.

8.2 Pathways to decarbonizing heat

Making the transition from gas domestic heating to low-carbon heating, such as heat pumps, requires our homes to be low-carbon ready (for example, insulated and double-glazed). Both consumers and the supply chain have to be willing and able to take the necessary steps and have a reason for doing so. A suggested pathway to decarbonizing heat is shown in Figure 8.1.

Following this pathway will require several measures, which could include the availability and affordability of measures to increase energy efficiency of homes, raising awareness of the options available, the availability and affordability of both low-carbon heating systems and energy supplies and the need to act.

Section 8 – Future perspectives: Heat as a Service

Figure 8.1 A suggested pathway to decarbonizing heat

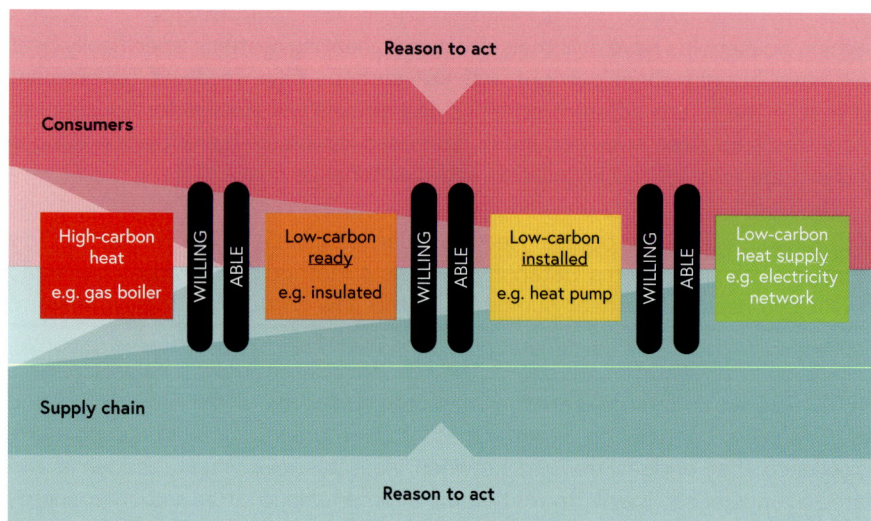

HaaS provides support to purchase or rent a new heating system so consumers can spread the cost of installing a low-carbon heating device. This does not address the willingness of either parties to install them, or consumers' willingness or ability to prepare their homes. HaaS has an opportunity to support more of the steps, and as evidence emerges of HaaS becoming commercialized from its projects with industry players, this shows that HaaS is a strong candidate to enable decarbonization of heat in homes.

It is important to understand why heating is important to people. Innovative new products and services need to understand what customers value and how to offer it to them.

Before considering what needs to change, it is important to recognize that people are not familiar with how they use heat at home. The government's Smart System and Heat (SSH) programme, delivered by the Energy Technologies Institute and later the ESC, carried out a review of consumers' preferences on heat and how it is used at home. This research highlights that about one-third of people think it is important to use their heating in terms of comfort. Another third are not interested in heating or what it is needed for, and the final third are cost-orientated and take significant control of their heating to conserve what they use[14].

The ESC's research also confirmed that consumers have problems related to heating at home: two-thirds of people have issues with damp or draughts, and many are unable to control their heating.

In the ESC's small-scale trials and advanced heating controls helped participants feel more benefit from the heat they bought. It was convenient, made people feel more in control, helped them understand how they like to be warm and what was possible in their home. People made use of choosing how warm to be in individual rooms, and when, and valued this experience[15]. They also thought it could be less wasteful by easily setting rooms to the temperatures and times they need, and subsequent analysis and modelling showed that advanced controls did save energy.

Advanced control of heat pump systems could be important in enabling low-carbon heat, by helping provide people with the heat and hot water experience they want, at a price they can afford. Heat

---

[14]Lipson, M An ETI Insights report - How can people get the heat they want at home, without the carbon? Energy Technologies Institute, Loughborough, 2018
[15]Haslett, A An ETI Insights report – The journey to smarter heat Energy Technologies Institute, Loughborough, 2019

pumps (with hot water tanks) can provide and store heat more flexibly, complementing the volatility of renewable generation. Combine this with advanced controls and people will better be able to plan their heating and buy it at times (and prices) that they prefer.

The idea of a service-led route to low-carbon heat draws from previous evidence outlined earlier that consumers care more about their experiences of using heat than how they are delivered. Offerings that specify heating outcomes are therefore seen to give consumers confidence that they will still be able to achieve the level of service they want from a low-carbon alternative. Their energy service provider will have a commercial incentive to learn how to deliver that service using as little energy as possible and, with an appropriate policy in place, with as little carbon as possible. The ESC developed a trial to demonstrate how consumers could buy outcomes and the heat experience they value through HaaS.

It was delivered through the ESC's Living Lab, Phase 2 of the government-funded SSH programme[16]. The Living Lab is a collection of over 100 households across the country that continues to expand and evolve. These gas boiler homes had the benefit of zonal heating control, and most households were active participants in the trial.

The trial was designed to explore how people can start to buy heat differently, and what is needed to do this. At present, consumers pay for their heating according to how much fuel they use. The SSH programme explored the idea that consumers could pay, instead, for the experience they get from their heating. Earlier research from the ESC suggested that consumers would find it more valuable to pay for the heat experience they like, rather than units of fuel as they do today[17]. These experiences could be offered and delivered through different energy services.

8.3   Heat Plans

The research introduced the Heat Plan concept (the ESC's version of HaaS) to consumers, and the transition of consumers paying for a set of desired outcomes rather than kWh consumed. This transformation would be similar to the mobile telecommunications industry, where consumers used to pay per minute, but now pay for talk time, data and text service contracts.

Heat Plans allow consumers to receive the heating they want, while paying a predictable fixed price for a specific outcome, such as an hour of heating. This is different from the current system where the cost of heating goes up and down for a number of reasons, such as due to the external temperature, amount of wind or internal doors being left open.

Heat Plans are bespoke as each household's heating needs are different and each home has different characteristics in the way it heats up and cools down. At the heart of the ESC's Heat Plan is the zonal smart heating system. The trial used a tailor-made control platform, but the sensors used were no different to those available in smart heating controls today. Using algorithms with data from the controls gives a valuable understanding of each home's building physics, and how people in the home like to be warm (see Figure 8.2).

There are many potential ways to construct a Heat Plan. The ESC created the concept of Heat Plans based on a 'warm hour', that is, any hour in a day where heating is requested in any room in a participant's home.

The trial included three different types of Heat Plan (FixedTime, FlexiTime or Unlimited), to introduce the concept that consumers can select and buy one of a range of different offers that might suit them. As the trial advanced, controls were developed by the ESC and the Heat Plans were offered to the trial participants on the smart control's app. The app was designed to mimic a virtual marketplace of different Heat Plan offers.

---

[16]https://es.catapult.org.uk/impact/projects/smart-systems-and-heat/smart-systems-and-heat-phase-2/
[17]Batterbee, J *Domestic Energy Services* Energy Technologies Institute, Loughborough, 2018

Section 8 – Future perspectives: Heat as a Service

Figure 8.2   An example of the zonal smart heating controls in the ESC's Living Lab trial

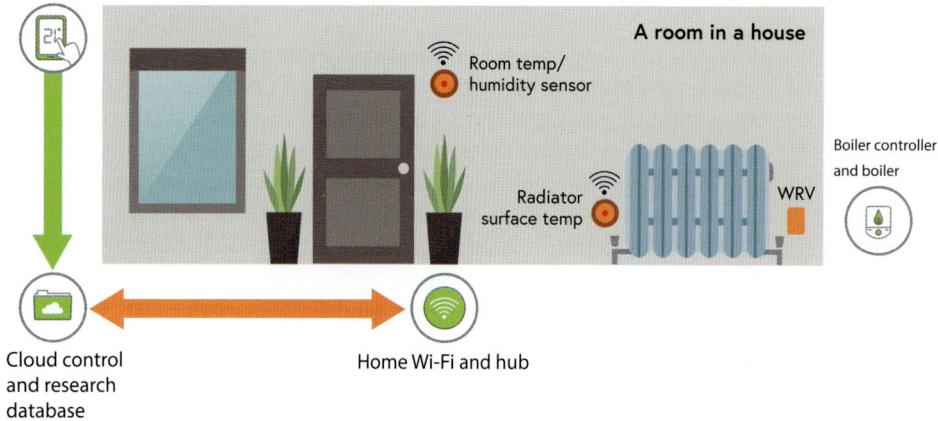

A Heat Plan asks consumers to pay for a certain number of warm hours each week as the basis for their plan. The different plans provide a specified warm-hour allowance and consumers were offered the choice to use additional warm hours.

Figure 8.3 shows:

(a) the number of warm hours included in the Heat Plan;

(b) whether the plan would allow the participant to alter their heat schedule and, if so, how that would work;

(c) the price of their plan each week;

(d) a breakdown of the cost of the plan in terms of pence per warm hour; and

(e) how much an extra warm hour would cost should the participant like to use more hours than were on their plan.

Figure 8.3   Example Heat Plan terminology

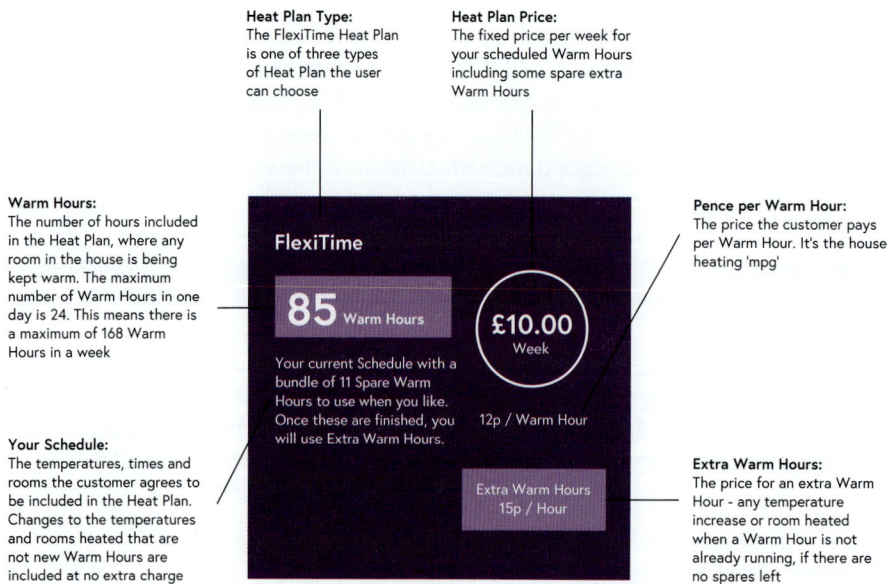

© The Institution of Engineering and Technology

Section 8 – Future perspectives: Heat as a Service

The ESC acted as the energy service provider for this trial. To make the Heat Plans seem realistic, participants were asked to imagine that they were dealing with a new energy service provider. Participants did not need to switch energy supplier as part of the initial trial. They were promised they would pay no more than the cost of their plan. If they used more gas, they were reimbursed the cost of that gas at the end of the trial. There was confidence to participants because many wanted to discuss the plan before committing. They clearly felt it was a real financial transaction. Later commercial trials delivered in collaboration with energy suppliers did switch participants, and the price of the Heat Plan (including any extras) was billed to the customer and paid directly to the supplier paying for kWh. This was only relevant for their electricity supply.

Figure 8.4 shows the different Heat Plans that were available.

Figure 8.4 The different Heat Plans available in the Heat Plan trial

8.3.1 FixedTime

The FixedTime plan was the least expensive and most restrictive Heat Plan and stipulated that:

(a) participants paid a lower, fixed price for the heating schedule they had set up at the time of the plan offers;
(b) participants were not able to change the schedule and were charged for any extra warm hours they used outside of their current schedule;
(c) participants could make changes to the temperature on demand, and it would have immediate effect; and
(d) the plan had a fair use policy applied to temperatures that could be requested and number of rooms heated.

Based on previous research it was assumed that this plan would appeal to participants who were more concerned about the cost of their heating than about being comfortable all the time. In particular, the assumption was that this plan would appeal to those who wanted to pay for a basic minimum amount of heat, had a predictable schedule for when they would be in and out of the house and would be unlikely to want to make many changes to their heating schedule.

Although other plans could have been designed for this type of consumer, the FixedTime plan was relatively easy to implement as it simply involved preventing any changes to the schedule once the plan had been accepted.

8.3.2 FlexiTime

FlexiTime was the mid-priced Heat Plan, allowing participants to change their schedule and stipulated that:

(a) a bundle of 'spare' warm hours could be used whenever participants liked, which offered greater flexibility for those whose needs changed over time;

(b) participants could change the temperature of each room they heated, subject to the fair use policy (see below);

(c) participants could also choose to heat a room they did not normally heat without incurring further costs, as long as it was within an existing heating period (subject to the fair use policy); and

(d) participants could also buy extra warm hours, if they needed more than the spare warm hours their plan allowed.

8.3.3 Unlimited

There were no restrictions with the Unlimited plan. Participants paid a higher fixed price for an unlimited number of warm hours to use as they liked. They could be as warm as they wanted.

It was possible through this plan to test a common pricing strategy, based on attracting customers who will pay a bit more for the best. This plan also allowed to test whether people would take the opportunity to use an excessive amount of heat – a common industry reservation related to energy services.

Fair use policy. Participants who chose the FixedTime or FlexiTime plans were subject to a 'fair use' clause. Fair use stipulated that if participants began to use their heating in a very different way (for example, frequently leaving windows open or using uncharacteristically high temperatures) they would be notified that they were exceeding the terms of their plan. Unlimited Heat Plans had no limits, for simplicity and to understand how people would use such a service, fair use did not apply.

8.4 User feedback

Approximately half those offered a Heat Plan during the Living Lab trial bought one. Of the Heat Plans available, the most popular was the FlexiTime plan.

The reason people gave for preferring this plan was that it allowed them to make changes in response to their changing needs or unpredictable schedules. The inclusion of spare warm hours also meant they were in a better position to adjust the number of warm hours as their requirements changed throughout the year.

The plan agreement allowed the researchers to cancel the Heat Plan if participants exceeded their fair use, although this was not necessary at any point.

Some participants reflected that the FlexiTime plan, while attractive, seemed 'indulgent'. Common factors that led participants to choose the FixedTime plan were the higher cost of the FlexiTime plan or the perception that the offered level of flexibility was not required.

The Unlimited plan was generally dismissed as not being necessary (although tempting to some). Of those who took plans, 83 % indicated that they felt the price quoted was 'fair' or, in some cases, 'cheap'.

This indicates that cost is not a priority for all, and some may be receptive to charges that are more than their current gas bill levels. The average increase that was accepted across all participants was £3.80 a week. This is in addition to the suppliers' margin, which was already included in their bills. The mean weekly cost of an accepted plan was £13.51, which is 39 % higher than the average weekly estimated cost of £9.71 for gas heating.

While some people accepted a plan because they felt it was comparable or cheaper to their current bill, some accepted plans that they understood were higher than their current bill. This indicates that some participants were receptive to paying more for a better service.

Section 8 – Future perspectives: Heat as a Service

Other participants indicated that they did not know their current bill level. The higher price of the plan was not necessarily clear to them, but they still felt that the offer was attractive.

It was important to test how likely participants would be to convert from a traditional gas boiler to a low-carbon alternative when their current boiler was due to be replaced. After all, this is one of the key assumptions behind HaaS being an enabler for the pathway to decarbonization. Participants who had chosen to take up a Heat Plan were asked if they would convert, with an assurance that the same heating experience as they had enjoyed during the trial would be maintained. The response from participants was that:

(a) more than half said that they would be open to the idea;

(b) a quarter were unsure, although they were receptive to receiving more information about low-carbon alternatives, and said they would be happy to adopt low-carbon equipment if their concerns and questions were addressed; and

(c) only a minority of people were completely closed off to the idea of moving away from a gas boiler, with only 16 % of those asked indicating that they would be unlikely to give low-carbon alternatives consideration.

This means that 84 % of the households that were asked about this subject could potentially convert to using a low-carbon alternative to their boiler.

The ESC also carried out a quantitative segmentation with over 2,000 consumers. As part of this, when asked about peoples' openness to low-carbon heating without introducing HaaS or the reassurance they would get the same heating experience that they do today, only about a third were interested[18]. Comparing this to the proportion of Living Lab participants that were interested in converting to low-carbon heating with the familiarity of HaaS (84 %) implies that buying HaaS could increase acceptance of upgrades to low-carbon heating.

8.5   Fuel poverty

About 2.5 million households in England live in fuel poverty –meaning they may not be able to afford the energy they need[19]. The UK government has spent approximately £3 billion a year over the past decade trying to alleviate fuel poverty, but it is still a key issue. There is also the additional burden placed on the NHS from those living in cold homes, which is reported to cost the NHS over £1 billion a year[20].

People at risk of fuel poverty do not just want to minimize what they spend, they want to be able to manage how much they spend whilst getting the heat they want[21]. What they want or need from their heating varies, as much as it varies for everyone else. Understanding how much it will cost them to get the heat they need can help them understand what they can afford.

HaaS could simplify people's experience of heating, with fixed price or simpler outcome-based tariffs giving them more control over their bills. The ESC's Heat Plans for example included options for participants to buy 'pay-as-you-go' warm hours. Instead of an allowance of warm hours each week, pay-as-you-go participants were charged for each warm hour they used, and were able to see how much it was costing them in real-time.

---

[18]D21 HESG Trial: System Test Reports and Trial Conclusions

[19]Department for Business Energy and Industrial Strategy (BEIS) *Annual Fuel Poverty Statistics Report 2020* BEIS, London, 2020. Available at: https://www.gov.uk/government/statistics/annual-fuel-poverty-statistics-report-2020

[20]Public Health England *Local action on health inequalities: Fuel poverty and cold home-related health problems* Public Health England, London, 2014

[21]Energy Services Catapult *Fuel Poverty in a Smart Energy World* Energy Services Catapult, Birmingham, 2020. Available at: https://es.catapult.org.uk/reports/fuel-poverty-in-a-smart-energy-world/

As well as helping the customer make more informed decisions on their comfort preferences based on cost, this data can also be used by, for example, public bodies, social housing providers and service providers to target subsidies and other financial support more effectively. This can be done through, for example, the data provided by smart control systems as part of HaaS delivery. This is the data needed to develop accurate estimates of the cost of warming homes so that it can be used to help target financial support to households at risk of fuel poverty. This data would need to be provided with consumer consent and there are mechanisms to ensure consumers have transparency on how their data is being used.

8.6   Next steps

Low-carbon heating system manufacturers, installers and above all service providers need to collaborate further on HaaS' delivery. There is an increasing awareness of market opportunities, and more organizations are exploring the options and there is also interest at a policy and regulatory level. Ofgem stated in its 2020 decarbonization action plan that innovative business models, such as Energy as a Service, will be needed to reach net zero[22]. They are open to helping organizations understand any necessary exemptions to deliver HaaS. The ESC's HaaS trials with commercial organizations suggest that exemptions may not be needed[23].

Revised policy and regulation to better support new business models, such as HaaS, will give service providers the reassurance and increased priority they need to accelerate HaaS' development. In the meantime, further trials and demonstrations of HaaS, such as those developed by the ESC, will help the industry learn and identify what is needed to deliver it themselves – and to support the route to net zero with low-carbon heat.

8.7   Summary

There will be many different ways in which electrified heat can be provided in a home ranging from the traditional purchase and operate to models, such as HaaS. HaaS offers opportunities for heat pump installers and manufacturers to build closer relationships with their customers and capture more of the supply chain. To progress, HaaS and similar hybrid models need low-carbon heating manufacturers, installers and above all service providers to collaborate further on its delivery.

It would be beneficial for installers and manufacturer to be taking part in trials, and to closely monitor government preparations on accelerating low-carbon heat in homes.

[22]Ofgem *Ofgem's decarbonisation action plan* Ofgem, London, 2020. Available at: https://www.ofgem.gov.uk/publications-and-updates/ofgem-s-decarbonisation-action-plan

[23]Energy Services Catapult *Bristol Energy as a service trial* Energy Services Catapult, Birmingham, 2020. Available at: https://es.catapult.org.uk/case-study/bristol-energy-heat-plan-trial/

Appendix A

Abbreviations

ACH	air changes per hour
AFDD	arc fault detection device
ASHP	air source heat pump
BECC	bioenergy with carbon capture and storage
BMS	building management system
CCC	Committee on Climate Change
CIBSE	Chartered Institution of Building Services Engineers
CO	carbon monoxide
CO_2	carbon dioxide
CoP	coefficient of performance
DBT	dry bulb temperature
DER	dwelling emission rate
DFEE	dwelling fabric energy efficiency
DHW	domestic hot water
DNO	distribution network operator
ENA	Energy Networks Association
EPC	energy performance certificate
ESC	Energy Systems Catapult
EV	electric vehicle
F-gas	fluorinated gas
GHG	greenhouse gas
GSHP	ground source heat pump
GWP	global warming potential
HaaS	Heat as a Service
HHRSH	high heat retention storage heating
HLP	heat loss parameter
HTC	heat transfer coefficients
HWHP	hot water heat pump
KVA	kilovolt amps
kWh	kilowatt hours
kW_{th}	thermal kilowatts
LED	light-emitting diode
m/s	metres per second
MCB	miniature circuit breaker devices
MCS	Microgeneration Certification Scheme

Appendix A – Abbreviations

MPAN	meter point administration number
MRT	mean radiant temperature
MVHR	mechanical ventilation heat recovery
MWh	megawatt hours
NOx	nitrogen oxide
NO$_2$	nitrogen dioxide
NPV	net present value
Ofgem	Office of Gas and Electricity Markets
P$_E$	evaporator pressure
P$_C$	compressor pressure
Pa	pascal
pa	per annum
PCM	phase change material
PHPP	Passivhaus Planning Package
PMV	predicted mean vote
PPD	predicted percentage dissatisfied
PV	photovoltaic
RCBO	residual-current circuit-breaker (with overcurrent protection)
RH	relative humidity
RHI	Renewable Heat Incentive scheme
SAP	standard assessment procedure
SCoP	seasonal coefficient of performance
SMETS	smart metering equipment technical specifications
SPF	seasonal performance factor
SSH	smart systems and heat programme
SSHEE	seasonal space heating energy efficiency
T$_E$	evaporator temperature
T$_C$	compressor temperature
TRV	thermostatic radiator valves
TWh	terrawatt hours
UFH	under floor heating
UKPN	UK Power Networks
V2G	vehicle to grid
V2H	vehicle to home
W/mK	watts per metre per degree Kelvin
W/m^{2}K	watts per square metre per degree Kelvin
Wh	watt hours
WSHP	water source heat pump

Appendix B

References

1. Department for Business, Energy and Industrial Strategy *Final UK greenhouse gas emissions national statistics: 1990 to 2018* BEIS, 2020. Available at: https://www.gov.uk/government/statistics/final-uk-greenhouse-gas-emissions-national-statistics 1990-to-2018

2. Department for Energy and Climate Change (DECC) *Digest of United Kingdom energy statistics* 60th anniversary DECC, London, 2009. Available at: https://www.gov.uk/government/statistics/digest-of-uk-energy-statistics-dukes-60th-anniversary

3. Department for Business, Energy and Industrial Strategy (BEIS) *Energy white paper: Powering our net zero future* BEIS, London, 2020. Available at www.gov.uk/government/publications/energy-white-paper-powering-our-net-zero-future

4. Met Office *UK Climate Projections* Available at: www.metoffice.gov.uk/research/approach/colla-boration/ukcp/index

5. Eyre, NJ, and Mallaburn, PS *Lessons from energy efficiency policy and programmes in the UK from 1973 to 2013* Energy Efficiency 7(1): 23–41. 2014

6. An operational overview of the GB gas national transmission system can be accessed at: www.nationalgrid.com/uk/gas-transmission/sites/gas/files/documents/Operational%20Overview%202018.pdf

7. *The Sixth Carbon Budget - The UK's path to Net Zero* December 2020. Available at: www.theccc.org.uk/publications

8. www.ofgem.gov.uk

9. *Domestic Building Services Compliance Guide* UK Government, London, 2013. Available at: https://assets.publishing.service.gov.uk/government/uploads/system/uploads/attachment_data/file/697525/DBSCG_secure.pdf

10. Chartered Institution of Building Services Engineers (CIBSE) *Guide A: Environmental design* CIBSE, London, 2015. Available at: https://www.cibse.org/getattachment/Knowledge/CIBSE-Guide/CIBSE-Guide-A-Environmental-Design-NEW-2015/Guide-A-presentation.pdf.aspx

11. IET *Code of Practice for Electric Vehicle Charging Equipment Installation, 4th Edition* IET, London, 2020

12. Ter-Gazarian, AG *Energy Storage for Power Systems, 3rd Edition* IET, London, 2020

13. IET *Code of Practice for Grid-connected Solar Photovoltaic Systems* IET, London, 2015

14. Lipson, M *An ETI Insights report - How can people get the heat they want at home, without the carbon?* Energy Technologies Institute, Loughborough, 2018

15. Haslett, A *An ETI Insights report – The journey to smarter heat* Energy Technologies Institute, Loughborough, 2019

16. https://es.catapult.org.uk/project/smart-systems-and-heat-phase-2/

Appendix B – References

17.   Batterbee, J *Domestic Energy Services* Energy Technologies Institute, Loughborough, 2018

18.   D21 *HESG Trial: System Test Reports and Trial Conclusions*

19.   Department for Business Energy and Industrial Strategy (BEIS) *Annual Fuel Poverty Statistics Report 2020* BEIS, London, 2020. Available at: https://www.gov.uk/government/statistics/annual-fuel-poverty-statistics-report-2020

20.   Public Health England *Local action on health inequalities: Fuel poverty and cold home-related health problems* Public Health England, London, 2014

21.   Energy Services Catapult *Fuel Poverty in a Smart Energy World* Energy Services Catapult, Birmingham, 2020. Available at: https://es.catapult.org.uk/report/fuel-poverty-in-a-smart-energy-world/

22.   Ofgem *Ofgem's decarbonisation action plan* Ofgem, London, 2020. Available at: https://www.ofgem.gov.uk/publications-and-updates/ofgem-s-decarbonisation-action-plan

23.   Energy Services Catapult *Bristol Energy as a service trial* Energy Services Catapult, Birmingham, 2020. Available at: https://es.catapult.org.uk/case-study/bristol-energy-heat-plan-trial/

▬ Appendix C

Standards

Approved Document Part F of the UK Building Regulations (2013) *Ventilation.*

Approved Document Part G of the UK Building Regulations (2016) *Sanitation, hot water safety and water efficiency.*

Approved Document Part L of the UK Building Regulations (2016) *Conservation of fuel and power.*

BSI (2018) BS 7671 *The IET Electrical Wiring Regulations.* London: BSI and IET.

BSI (2009) BS 5422 *Method for specifying thermal insulating materials for pipes, tanks, vessels, ductwork and equipment operating within the temperature range −40 °C to 1,700 °C.* London: BSI.

BS EN ISO 7730:2005 Ergonomics of the thermal environment. Analytical determination and interpretation of thermal comfort using calculation of the PMV and PPD indices and local thermal comfort criteria.

BSI (2005) BS EN 61000-3-12 (2005) *Electromagnetic compatibility (EMC). Limits. Limits for harmonic currents produced by equipment connected to public low-voltage systems with input current > 16A and ≤ 75A per phase.* London: BSI.

BSI (2008) BS EN 61000-3-3 *Electromagnetic compatibility (EMC). Limits. Limitation of voltage changes, voltage fluctuations and flicker in public low-voltage supply systems, for equipment with rated current ≤ 16 A per phase and not subject to conditional connection.* London: BSI.

BSI (2019) BS EN IEC 61000-3-11 *Electromagnetic compatibility (EMC). Limits. Limitation of voltage changes, voltage fluctuations and flicker in public low-voltage supply systems. Equipment with rated current ≤ 75 A and subject to conditional connection.* London: BSI.

BSI (2019) BS EN IEC 61000-3-2 *Electromagnetic compatibility (EMC). Limits. Limits for harmonic current emissions (equipment input current ≤ 16 A per phase).* London: BSI.

BSI (2019) PAS 2035/2030 *Retrofitting dwellings for improved energy efficiency. Specification and guidance.* London: BSI.

European Standards (2019) EN 14511 Part 1 *Air conditioners, liquid chilling packages and heat pumps for space heating and cooling and process chillers, with electrically driven compressors.* Berlin: Deutsches Institut für Normung.

Fabric Energy Efficiency Standard (FEES).

Microgeneration Certification Scheme (2019) 020 *Planning Standards for permitted development installations of wind turbines and air source heat pumps on domestic premises.*

Microgeneration Installation Standard (MIS) MCS 3005 (2017) *Requirements for MCS contractors undertaking the supply, design, installation, set to work, commissioning and handover of microgeneration heat pump systems* (MCS).

BSI (2015) BS 8558:2015 *Guide to the design, installation, testing and maintenance of services supplying water for domestic use within buildings and their curtilages. Complementary guidance to BS EN 806.* London: BSI.

Appendix D

Schematics

The following schematics, while not exhaustive, have been included to show the reader a range of different options they may find useful.

Figure D.1 ASHP, DHW thermal store and heating schematic

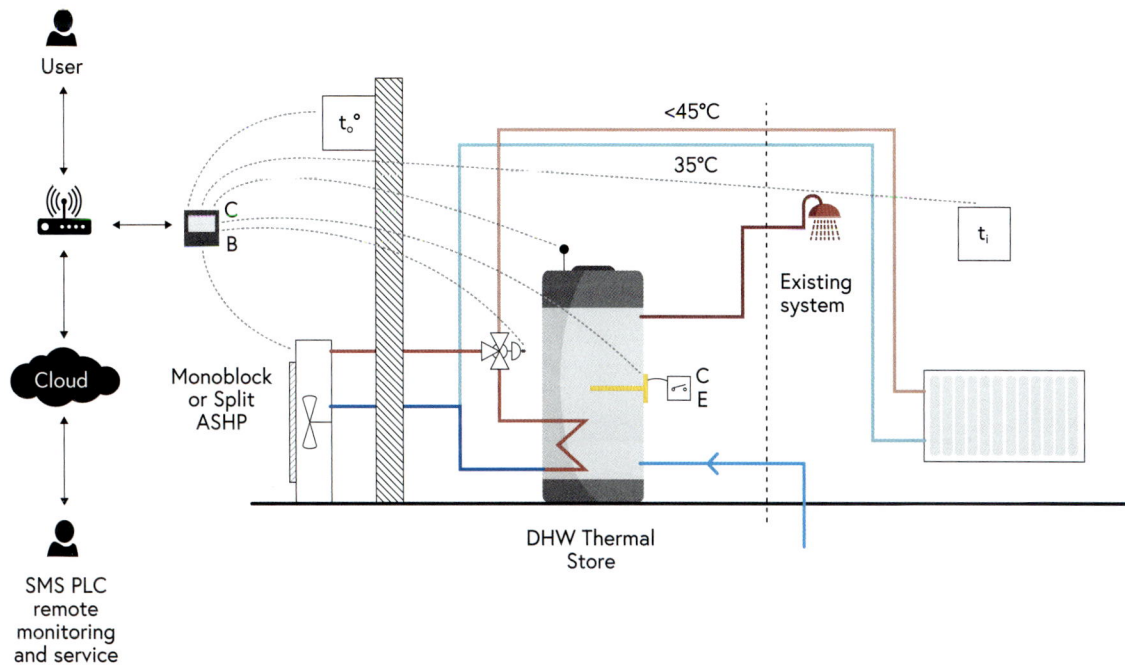

Notes:

- New monobloc or split ASHP for heating and DHW supply.
- Recommended heating FLOW temperature is $\leq 45\,°C$, weather compensation flow temperature control philosophy.
- Stored DHW temperature is maximum $55\,°C$ DHW thermal store is fitted with a 3 kW electric immersion heater for weekly Legionella cycle and back-up heating source.
- Existing radiator may need to be upgraded for the lower flow temperature.

Appendix D – Schematics

Figure D.2 ASHP, Pre-plumbed DHW thermal store and heating schematic

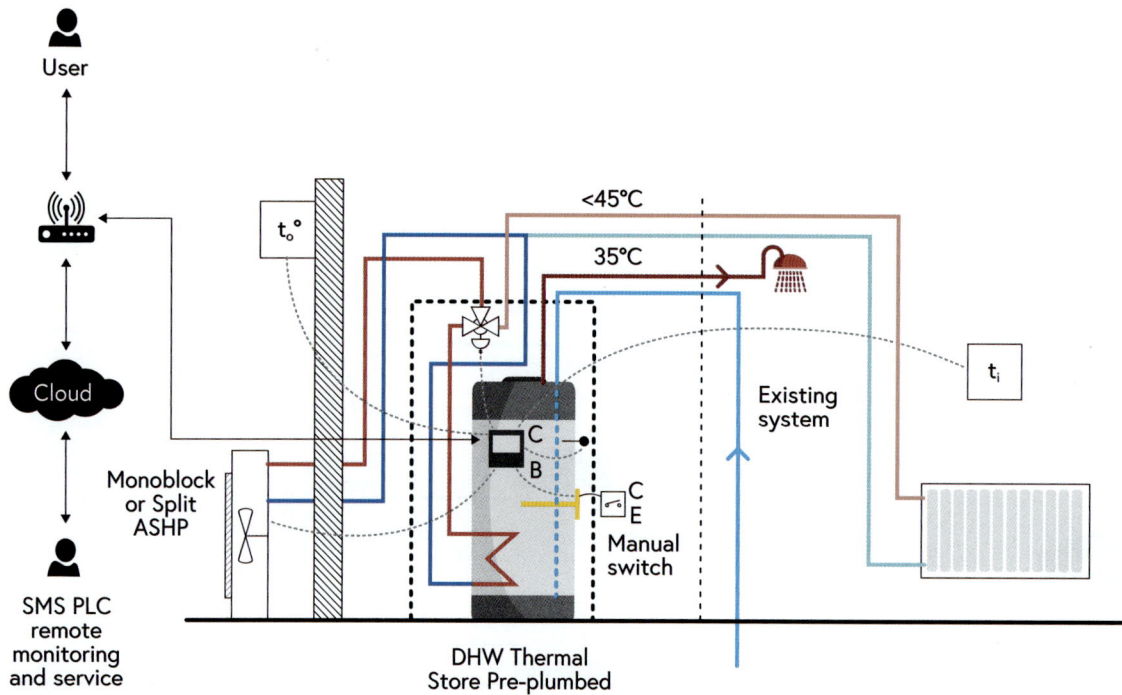

Notes:

- New monobloc or split ASHP for heating and DHW supply.
- Recommended heating FLOW temperature is $\leq 45\,°C$, weather compensation flow temperature control philosophy.
- Pre-fitted (plug and play) thermal store shall be used to reduce space requirement and installation time.
- Stored DHW temperature is maximum $55\,°C$ DHW thermal store is fitted with a 3 kW electric immersion heater for weekly Legionella cycle and back-up heating source.
- Existing radiator may need to be upgraded for the lower flow temperature.

Appendix D – Schematics

Figure D.3   ASHP, DHW thermal store and underfloor heating schematic

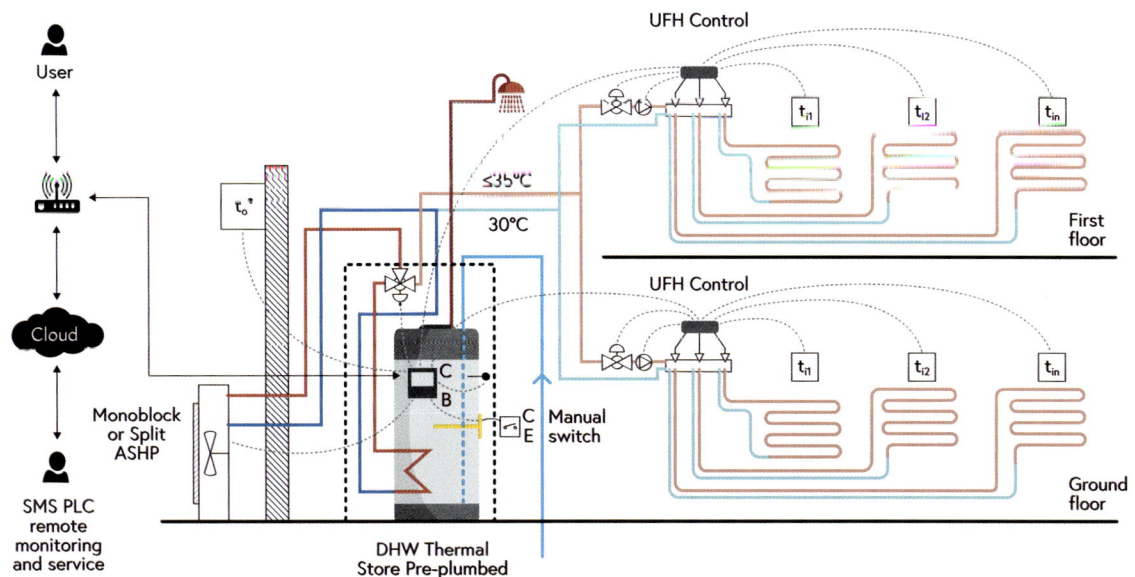

Notes:

- New monobloc or split ASHP for heating and DHW supply.
- Recommended heating FLOW temperature is $\leq 35\,°C$, weather compensation flow temperature control philosophy.
- Pre-fitted (plug and play) thermal store shall be used to reduce space requirement and installation time
- Stored DHW temperature is maximum $55\,°C$ DHW thermal store is fitted with a 3 kW electric immersion heater for weekly Legionella cycle and back-up heating source.
- New underfloor heating for space heating.
- Each underfloor heating zone has own zone thermostat to achieve the best efficiency.

Appendix D – Schematics

Figure D.4 ASHP, DHW thermal store, system boiler and heating schematic

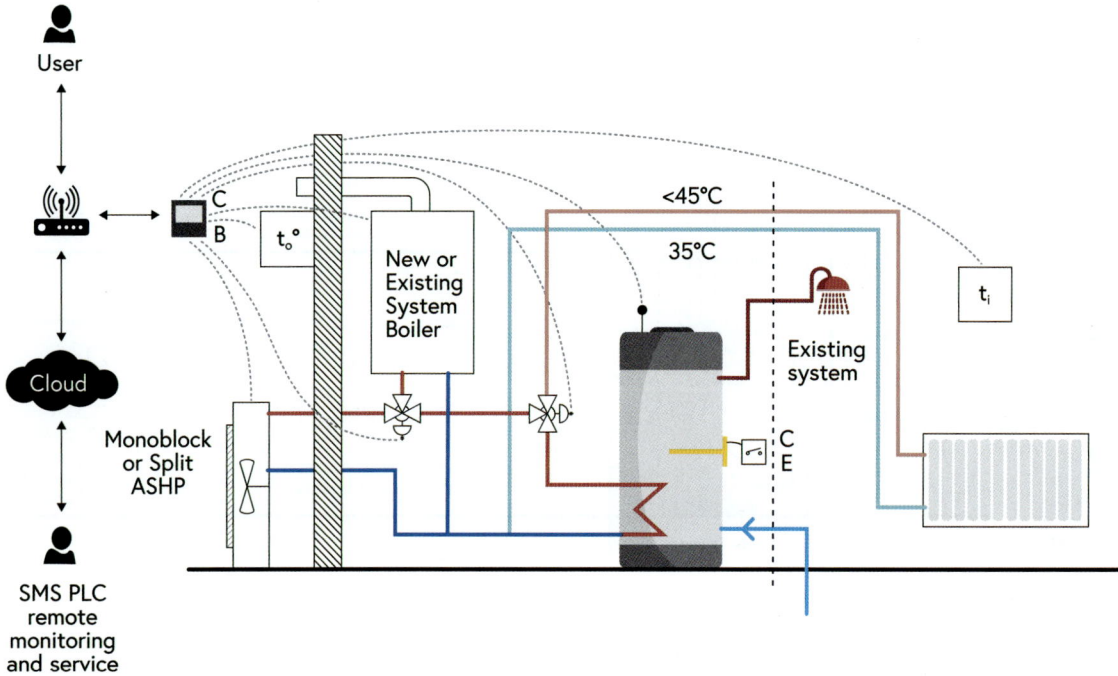

Notes:

- New monobloc ASHP for heating and DHW supply.
- Recommended heating flow temperature can be adjusted by the ASHP and gas system boiler according with the requirement.
- Stored DHW temperature is maximum 55 °C DHW thermal store is fitted with a 3 kW electric immersion heater for back-up heating source.
- Weekly Legionella cycle can be done by the gas system boiler.
- Existing radiator may need to be upgraded for the lower flow temperature.

Appendix D – Schematics

Figure D.5 ASHP, gas combi boiler, DHW and heating schematic

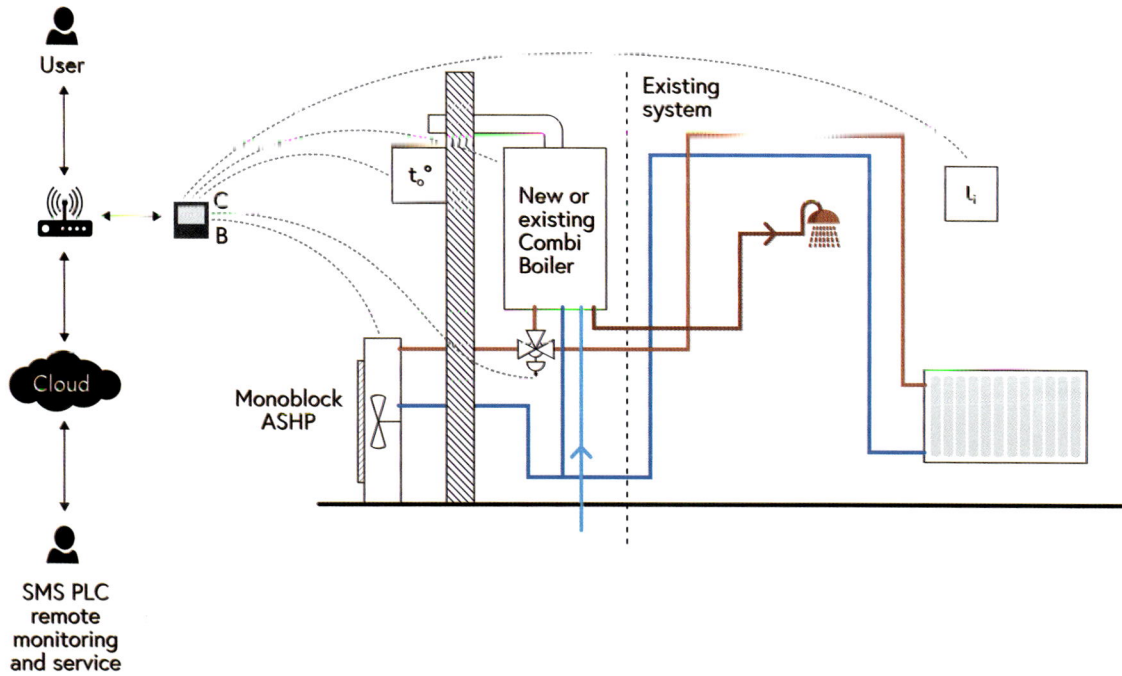

Notes:

- New monobloc ASHP for heating supply only.
- Heating flow temperature can be adjusted by the ASHP and gas system boiler according to the requirement.
- DHW supplied by gas combi boiler.
- Existing radiator may need to be upgraded for the lower flow temperature.

Appendix D – Schematics

Figure D.6 Future ASHP connected to existing gas combi boiler, DHW and heating schematic

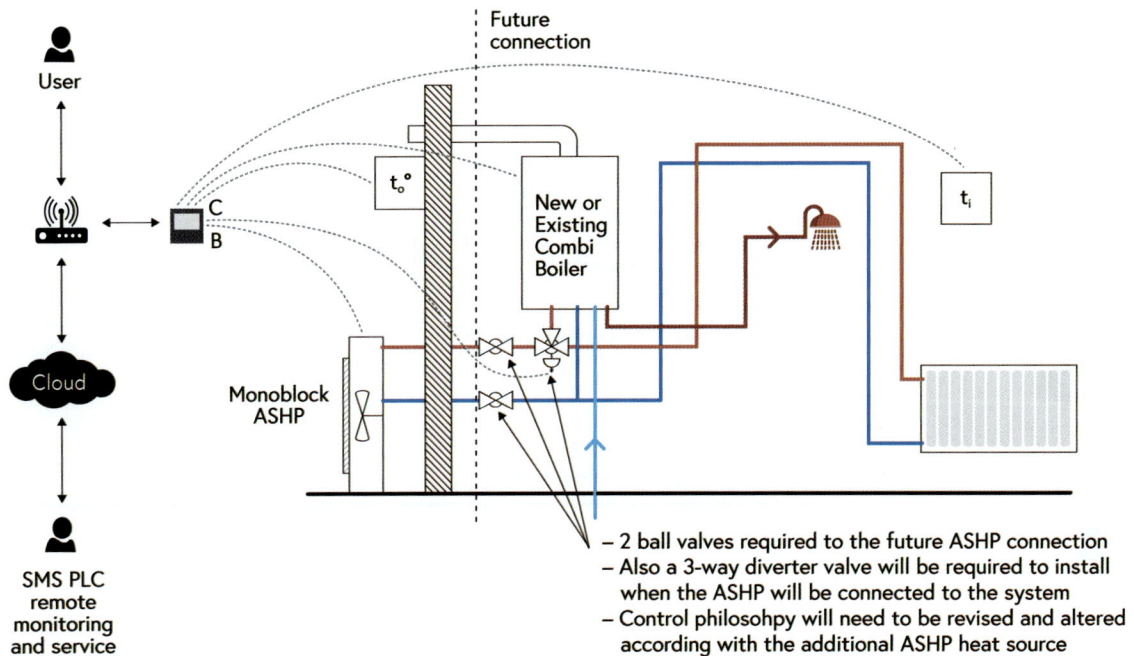

- 2 ball valves required to the future ASHP connection
- Also a 3-way diverter valve will be required to install when the ASHP will be connected to the system
- Control philosohpy will need to be revised and altered according with the additional ASHP heat source

Notes:

-   New monobloc ASHP for heating supply only.
-   Heating flow temperature can be adjusted by the ASHP and gas system boiler according to the requirement.
-   DHW supplied by gas combi boiler.
-   Existing radiator may need to be upgraded for the lower flow temperature.

Appendix D – Schematics

Figure D.7 ASHP providing heating and gas combi boiler providing DHW schematic

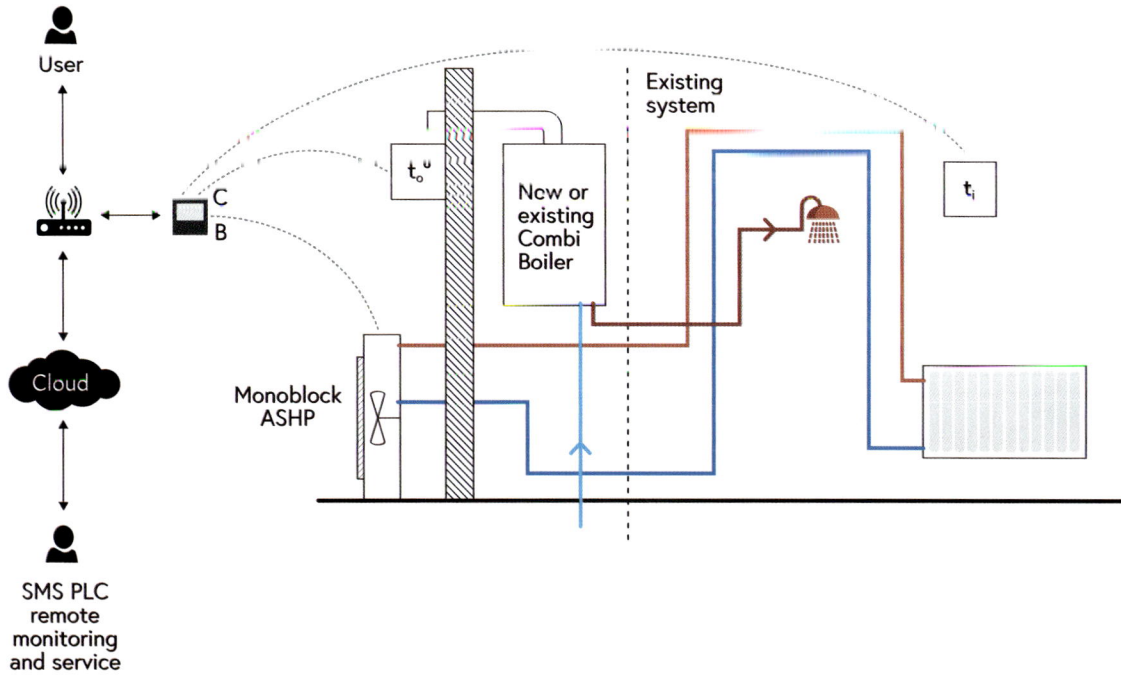

Notes:

– New monobloc ASHP for heating supply only.
– Heating flow temperature can be adjusted by the ASHP according to the requirement.
– DHW supplied by gas combi boiler.
– Existing radiator may need to be upgraded for the lower flow temperature.

Appendix D – Schematics

Figure D.8   ASHP providing heating and direct electric water heating providing DHW schematic

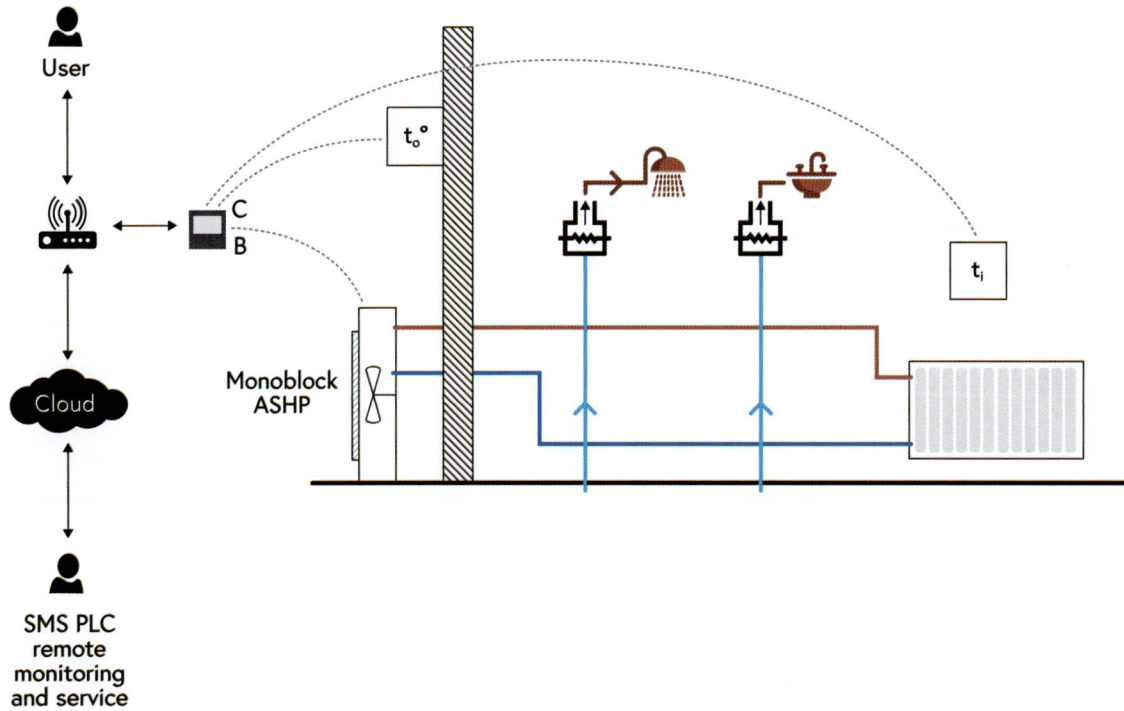

Notes:

- New monobloc ASHP for heating supply only.
- Heating flow temperature can be adjusted by the ASHP according to the requirement.
- DHW supplied by electric under sink water heater and electric shower.
- Existing radiator may need to be upgraded for the lower flow temperature.

Appendix D – Schematics

Figure D.9   DHW supplied by electric under sink water heater and electric shower

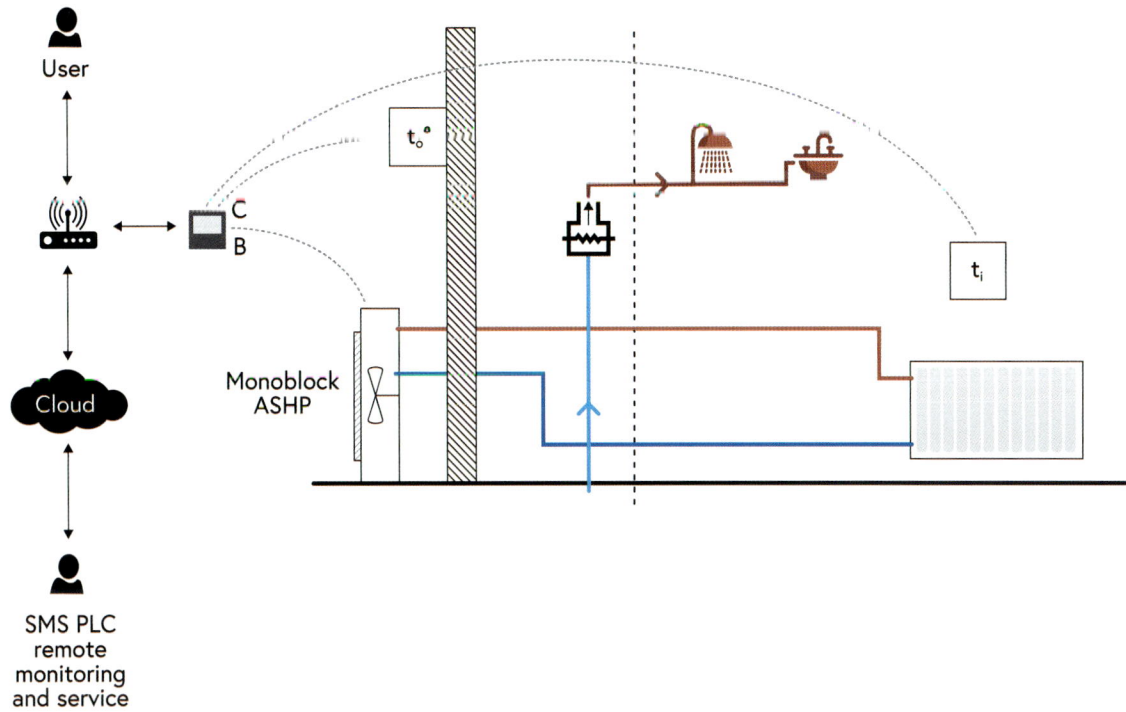

Notes:

- New monobloc ASHP for heating supply only.
- Heating flow temperature can be adjusted by the ASHP according to the requirement.
- DHW supplied by electric flow boiler water heater.
- Existing radiator may need to be upgraded for the lower flow temperature.

▬ Index

Index

NOTES

NOTES